Lectures on Reflexive game Theory

Vladimir A. Lefebvre

Leaf & Oaks Publishers

Los Angeles

First publication Cogito-Centre, 2009 (in Russian)
Translation by Victorina D. Lefebvre

This book describes an innovative approach to reflexive game theory.
The applications of this theory include predicting and influencing
choices made by individual subjects belonging to groups that have their
own collective goals and interests. The correlation between a subject's
individual interests and those of the group is informed by the anti-
selfishness principle: a subject belonging to a group, in pursuing his or
her own interests, may not cause harm to the interests of the group as a
whole. This principle is as foundational to reflexive game theory as the
principle of guaranteed results in classical game theory.

Numerous examples are given of reflexive game-theoretical analysis in
personal relations, politics, international relations, military decision-
making, and jurisprudence.

There is also an Appendix containing problems and exercises relevant to
the material described.

Key words: reflexive games, reflexive control, decision making, military
decisions, social structures, human choice, international relationships

ISBN 978-0-578-06594-6

Contents

Introduction

These lectures do not follow in the tradition of classical game theory. They do not employ concepts such as 'strategy', 'utility', 'payoff matrix', 'guaranteed result', and 'equilibrium point'. The reflexive game theory is intended to address problems different from those addressed by classical game theory. Its goal is to predict the individual choice made by a subject belonging to a group and outline the possibility for controlling this choice. We call this *reflexive control*. The term 'subject' refers to single individuals or to various types of organizations: political parties, military units, states, and even civilizations. Relations between the interests of a group and those of individual subjects are regulated by a rule called *the anti-selfishness principle*: a subject in a group while pursuing one's own goal may not cause harm to the group as a whole. This principle is as important in reflexive game theory as is the principle of guaranteed results in classical game theory.

The anti-selfishness principle renders actions unacceptable if they are advantageous to the subject but harmful for the group to which the subject belongs. It does not, however, forbid antisocial actions, if the subject does not undertake such actions in pursuing of his own interests. In this way, the subject's unselfishness may justify actions that cause harm to a group or society. Note that which actions are preferable for a subject and for a group are not determined in advance, but rather generated formally by the model itself.

An essential difference from classical game theory is that reflexive game theory makes special assumptions concerning the mental mechanism of choice. We assume that the subject possesses a partially ordered set of self-images, that is, the subject has several images of the self, each of which may have images of the self, etc. The hierarchy of images is depicted by a special formula that we call *diagonal form*. Let us emphasize that the hierarchy of images is not arbitrary. It is always finite and predetermined by the graph of relationships among subjects. This is an important point. People have known for many centuries that the mental domain can be described by recursive chains of the type, "he knows that he knows that he knows . . . " Such chains have not been used in the models of human cognition, however, because there was no "stopping rule." That such a rule exists in reflexive game theory makes the theory viable.

Diagonal form defines a mental procedure of choice and, at the same time, the mathematical function describing that choice. Thus, having written a diagonal form according to empirical data, we automatically obtain the subject's choice function. One of the values in the diagonal form is interpreted as the subject's intention. Subsequently, we assume that the subject is intentional, which means that he has only such intentions as can become reality; thus, intentions are not assigned in advance. The intentional subject corresponds to an equation whose solution is interpreted as an alternative that can be chosen by the subject. The case where the equation has no solutions is interpreted as the subject's inability to make a decision under the given circumstances. The equation may have several solutions; each of them is considered to be a possible choice. Finally, there may be a case that any alternative can be the solution; we assume then that the subject has freedom of choice, i.e., the group does not impose constraints on his decision.

In this book, we provide examples applying reflexive game theory in the areas of personal relations, social life, politics, international relationships, military decision making, and law.

Is it possible to test objectively this theory? The principle of falsification proposed by Karl Popper for the natural sciences can hardly be used in this case. General social-psychological theories cannot be considered wrong due to small numbers of failures, because their field of applications is defined only vaguely. The idea formulated by Donald Campbell (1997) fits better: in science, there is a principle of natural selection by which only those theories that are interesting for researchers survive. Reflexive game theory as well as classical game theory cannot be rejected due to numbers of failed predictions. Its fate depends on whether specialists use it and for how long.

The first step in constructing reflexive game theory was made more than forty years ago. The recursive chains, "I know that he knows that I know . . . " have been to found a model of the subject, with reference to the term "reflexive games" (Lefebvre, 1965, 1966, 1967). Then a special formal apparatus was developed for modeling human choice (Lefebvre, 1982), which applied the reflexive game theory to the analysis of specific situations (Lefebvre, 2001, 2007).

Important contributions to the theoretical understanding of reflexion have been made by numerous researchers. Tatiana Taran (1998, 2001) constructed a multivalued Boolean model for choosing social norms. Vladimir Krylov (2000) studied problems related to axiomatics of reflexive models. Yuli Schreider (1999) considered continuous-valued logics as languages of reflexion. Pavel Baranov and Vladimir Lepsky developed a formal model of the subject with reflexion and inner value (see Lefebvre, Baranov, Lepsky, 1969). Anatoly Trudolubov (1972) created a reflexive game model using dependency nets. Tim Kaiser and Stefan Schmidt (2008) found relations between reflexive game theory and the theory of functors and categories. There were also two attempts to combine reflexive games with classical game theory. Novikov and Chkhartashvili (2003) included reflexive games into the formalism of classical game theory; Lefebvre (2001) included classical game theory within reflexive game theory. Future study will show whether such connections may be productive.

The book includes an Appendix. It contains problems and exercises relevant to the material described.

I am grateful to Viacheslav Filimonov, Tim Kaiser, Stefan Schmidt, Jonathan Farley, and Harold Baker for their advice and corrections. Invaluable help was given to me by Victorina Lefebvre, with whom I discussed my plans and the main ideas of the book. In addition, she prepared these lectures for publication and made many significant comments in the course of work. Without her participation this book would never have been completed.

Chapter 1

Sets, Boolean algebras, exponential formulae and equations

This chapter sets out the fundamental concepts of set theory and Boolean algebra that will be used to construct a formal theory of reflexive games.

1.1. Sets

A *set* is a collection of distinguishable elements of any nature. An abstract object that does not contain any element is also called a set - an *empty set*. Let us consider a non-empty set that we call *universal* and designate as 1. The set of all subsets of the universal set 1, including the empty set, 0, will be designated as M. We assume that every set includes itself as a subset. A set containing the elements $\alpha, \beta, \gamma, \ldots$ is denoted as $\{\alpha, \beta, \gamma, \ldots\}$.

The expression $A \supseteq B$, where $A \in M$ and $B \in M$, means that B is a subset of A; B may be the set A itself. The expression $A \supset B$ means that B is a proper subset of A, $B \neq A$. We will designate the union of two sets as $+$, and their intersection as $\cdot$.

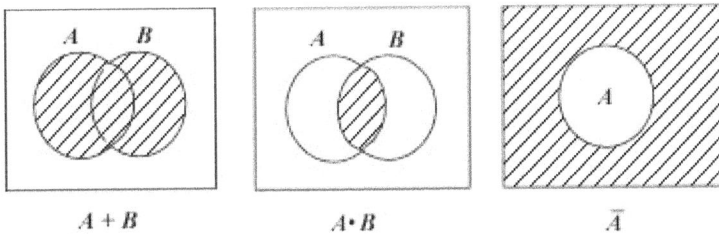

$$A + B \qquad A \cdot B \qquad \bar{A}$$

Fig.1.1.1. Venn diagrams.
The square in the figure designates the universal set;
its shaded part is the result of the given operation.

A line above a letter designates the unary operation of complement to the given set, that is, $\overline{A}$ is the set of elements of the universal set that are not included in A. To simplify working with sets we will use Venn diagrams. Figure 1.1.1 shows examples of union $A+B$, intersection $A \cdot B$, and complement $\overline{A}$.

To simplify notation we will write $B \cdot C$ as BC. We will interpret the expression $A+BC$ to mean that the intersection of B and C is first found and then combined with A. The following correlations hold for subsets of set M:

1. $A + A = A$	2. $A\,A = A$
3. $A + B = B + A$	4. $A\,B = B\,A$
5. $A + (B + C) = (A + B) + C$	6. $A\,(B\,C) = (A\,B)\,C$
7. $A\,(B + C) = AB + AC$	8. $A + BC = (A+B)\,(A+C)$
9. $A + B = \overline{\overline{A}\,\overline{B}}$	10. $A + 0 = A$
11. $A + 1 = 1$	12. $\overline{\overline{A}} = A$
13. $A + \overline{A} = 1$	14. $\overline{1} = 0$

1.2. Boolean algebras

Correlations 1-14 above coincide with axioms of Boolean algebra. Therefore, set M with the operations $+$, $\cdot$, $^{-}$ and the relation $\supseteq$ can be regarded as a Boolean algebra. If the universal set consists of one element, α, then the Boolean algebra consists of two elements: 1 – the universal set containing element α - and the empty set $\{\ \}$. If the universal set consists of two elements, α and β, then the Boolean algebra consists of four elements:

$$1=\{\alpha,\ \beta\},\ \{\alpha\},\ \{\beta\},\ 0=\{\ \}.$$

In general, if a universal set consists of k elements, the corresponding Boolean algebra consists of 2^k elements (power set).

It is convenient to represent Boolean algebras as lattices: edges correspond to relations of the form $A \supset B$, where B is an element located below A. Examples are given in Figures 1.2.1 and 1.2.2.

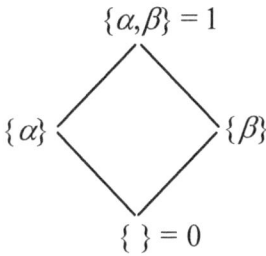

Fig.1.2.1. Boolean lattice corresponding to universal set of two elements, α and β

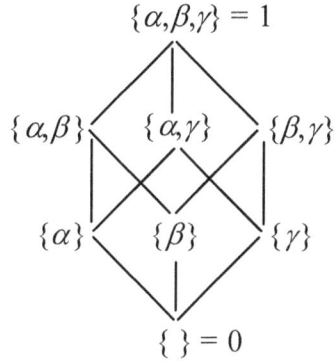

Fig.1.2.2. Boolean lattice corresponding to universal set of three elements, α, β and γ

From now on, in speaking of the set M, i.e., the set of all subsets of the universal set 1, we will keep in mind that M is a Boolean algebra with operations $+$, $\cdot$, $^{-}$ and relation $\supseteq$. It is possible to define functions on the set M that map groups of elements from M onto elements from M. For example,

$$f(a,b,c) = a + bc,$$

where $a,b,c \in M$. This function maps any three elements a,b,c onto $a + bc$.

1.3. Exponential formulae

The following equation plays an important role in our considerations:

$$\Phi(a,b) = a + \bar{b} . \qquad (1.3.1)$$

We will write it conventionally in exponential form

$$\Phi(a,b) = a^b \tag{1.3.2}$$

and accept the convention for multi-leveled exponents

$$a^{b^c} = a^{\left(b^c\right)}. \tag{1.3.3}$$

The following correlations hold:

1. $a^b a^c = a^{b+c}$ 2. $\left(a^b\right)^c = a^{bc}$

3. $(ab)^c = a^c b^c$ 3. $a^b + a^c = a^{bc}$

5. $(a+b)^c = a^c + b^c$ 6. $(a+b)^c = a^c + b$

7. $a^c + b = a + b^c$ 8. $a^{a+b} + b^{a+b} = 1$

9. $a^a = 1$ 10. $a^{ab} = 1$

11. $(a+b)^a = 1$ 12. $a^b + b^a = 1$

13. $a^0 = 1$ 14. $1^a = 1$

15. $a^1 = a$ 16. $a^{\bar{a}} = a$

17. $0^a = \bar{a}$

Every exponential expression can be transformed into linear notation. For example,

$$a^{b^{c+d}} = a^{b + \overline{c+d}} = a^{b + \overline{c}\overline{d}} = a + \overline{b + \overline{c}\overline{d}} = a + \overline{b}\ \overline{\overline{c}\overline{d}} = a + \overline{b}(c+d) \quad.$$

In many cases, it is not necessary to transform an expression into linear form; it may be easier to substitute the values of variables directly to an exponential equation. Let the formula

$$a^{b^{cd}},$$

be defined on set M of the universal set $\{\alpha,\beta,\gamma\}$. Set M consists of eight elements:

$$1 = \{\alpha, \beta, \gamma\}, \{\alpha, \beta\}, \{\alpha, \gamma\}, \{\beta, \gamma\}, \{\alpha\}, \{\beta\}, \{\gamma\}, \{\ \} = 0.$$

Let $a = \{\alpha\}$, $b = \{\alpha, \beta\}$, $c = \{\beta, \gamma\}$, $d = 1$. By substituting these values into the exponential form we obtain

$$\{\alpha\}^{\{\alpha,\beta\}^{\{\beta,\gamma\}1}} = \{\alpha\}^{\{\alpha,\beta\}+\overline{\{\beta,\gamma\}}} = \{\alpha\}^{\{\alpha,\beta\}+\{\alpha\}} =$$

$$= \{\alpha\}^{\{\alpha,\beta\}} = \{\alpha\} + \overline{\{\alpha,\beta\}} = \{\alpha\} + \{\gamma\} = \{\alpha,\gamma\}.$$

Computations of complex exponential formulae and universal sets with large numbers of elements can be done in a similar way.

1.4. Equations

Consider function

$$y = Ax + B\bar{x} , \tag{1.4.1}$$

where x, A, $B \in M$; A and B do not depend on x.

Statement 1.4.1. Equation

$$Ax + B\bar{x} = x \tag{1.4.2}$$

has a solution if and only if

$$A \supseteq B . \tag{1.4.3}$$

Proof. Let the equation have a solution. Then $Ax = x$ and $B\bar{x} = 0$. It follows from the first equality that $A \supseteq x$, and from the second that $x \supseteq B$, because if the complement to x does not intersect with B, then set B is within x. Therefore, $A \supseteq B$. Now let $A \supseteq B$. Choose x such that $A \supseteq x \supseteq B$. It is clear that $Ax = x$ and $B\bar{x} = 0$, since the intersection of B with the complement to x is empty. Therefore, x is a solution of (1.4.2) □

It follows from the preceding proof that every x from the interval $A \supseteq x \supseteq B$ is a solution of equation (1.4.2).

Examples. Consider the following equation:

$$Px + R = x. \tag{1.4.4}$$

Let us find whether (1.4.4) has a solution. To do so, we must represent it as (1.4.2). Since $x + \bar{x} = 1$, we can write (1.4.4) as

$$Px + R(x + \bar{x}) = x$$

and transform it into

$$(P + R)x + R\bar{x} = x .$$

In this equation, $P + R$ plays the role of A, and R plays the role of B, that is,

$$(P + R) \supseteq R .$$

Thus, (1.4.4) has at least one solution. It belongs to the interval

$$(P + R) \supseteq x \supseteq R .$$

Consider another equation:

$$P + R\bar{x} = x , \tag{1.4.5}$$

where $P \subset (P + R)$.

We transform its left side:

$$P(x + \bar{x}) + R\bar{x} = Px + (P + R)\bar{x} .$$

In this case,

$$A = P, \quad B = P + R .$$

We see that $A \subset B$, i.e., condition (1.4.3) is not met, which means that equation (1.4.5) does not have a solution.

Finally, consider an equation that takes its values from set M of the universal set $\{\alpha, \beta, \gamma\}$:

$$\{\alpha,\beta\}x + \{\beta,\gamma\}\overline{x} = x \ , \qquad (1.4.6)$$

$$A = \{\alpha,\beta\}, \ B = \{\beta,\gamma\} \ .$$

In this case, condition (1.4.3) is not met because B is not a subset of A; thus, equation (1.4.6) does not have a solution.

Equations can be written in an exponential form, as well. For example,

$$a^{b+c^{x}} = x \ . \qquad (1.4.7)$$

We transform the left part:

$$a^{b+c^{x}} = a^{b+c+\overline{x}} = a + \overline{b}\overline{c}x \ .$$

Now the equation has the form

$$a + \overline{b}\overline{c}x = x \ .$$

Let us represent the left side as $Ax + B\overline{x}$:

$$a + \overline{b}\overline{c}x = a(x + \overline{x}) + \overline{b}\overline{c}x = (a + \overline{b}\overline{c})x + a\overline{x} \ ,$$

$$A = a + \overline{b}\overline{c}, \ B = a \ .$$

We see that $A \supseteq B$, and the solutions of (1.4.7) are given by the inequalities

$$(a + \overline{b}\overline{c}) \supseteq x \supseteq a \ .$$

Chapter 2
Complete graphs with edges of two types

We assume that any two subjects within a group are either in a relationship of cooperation or in confrontation. We can represent a group as a graph, whose nodes correspond to subjects and whose edges correspond to the relations between them. Graphs in which every two nodes are linked are called *complete*. The edge connecting two nodes, a and b, is designated (a,b). Edges (a,b) and (b,a) are equivalent. A graph is called *elementary* if it consists of one node. The set of all edges of a non-elementary graph can be divided into two nonintersecting subsets, one of which may be empty. We will call these subsets *relations R and $\overline{R}$* . One of them is interpreted as cooperation, and the other as conflict. If $(a,b) \in R$, we say that a and b are linked by R, and write aRb. If $(a,b) \in \overline{R}$, then a and b are linked by $\overline{R}$, which is written $a\overline{R}b$. All definitions for R hold for $\overline{R}$. If two nodes, a and b, can be linked by a sequence of R-edges, we say that a and b are linked in R. If every node of graph A is linked with every node of graph B in R, we will write it ARB. In this case, we say that A and B are in relation R. Thus, relation R between nodes is transferred to a relation between graphs. Expression aRb will also designate the relation between elementary graphs consisting of nodes a and b, respectively. If graph G consists of subgraphs that are mutually in relation R, we say that G is *divided* into these subgraphs; we will write this down as $G = A_1RA_2R...RA_n$, where $A_1, A_2,...,A_n$ are subgraphs. Expression $B \subseteq A$ means that graph B is a subgraph of A: B's set of nodes is a subset of A's nodes, its edges are induced by A (if B is not an elementary graph), that is, every edge in B linking two of its nodes a and b

coincides with an edge in A linking these nodes. Expression $B \subset A$ means that B is a subgraph of A, but does not coincide with A. If $B \subseteq A$ and $C \subseteq A$, then $D = B \cup C$ means that D is a union of B's and C's sets of nodes with edges (if they exist) induced by A, and $E = B \cap C$ means that E is an intersection of B's and C's sets of nodes (if the intersection is not empty), with edges (if they exist) induced by A. Expression $a \in A$ means that a is a node of graph A, and $A_{(k)}$ means that graph A consists of k nodes. Expression $G–A$ denotes graph G, from which all A's nodes are taken away. Expression $<a,b,...>$ denotes a graph with nodes a, b, ... Further on we will consider complete and elementary graphs. The proofs of statements formulated in this Chapter and in the next one are given in Appendix 10 to the Lefebvre's book (2001).

2.1. Basic definitions

Definition 1. Graph G is *stratified* in R, if it can be represented as $G = ARB$. Graphs A and B are called *strata* of graph G in R.
Definition 2. Graph G is *totally stratified*, if each of its non-elementary subgraphs is stratified either in R or in $\overline{R}$.
Definition 3. If A is a stratum of G in R, and if A is not stratified in R, A is called a *minimal stratum* of G in R.

2.2. Theorem on total stratification

A graph consisting of four nodes linked both in R and in $\overline{R}$, will be denoted $S_{(4)}$. An example of such a graph is given in Fig. 2.2.1. Solid lines depict relation R, and broken ones depict relation $\overline{R}$. The graph is linked both by solid lines and broken ones.

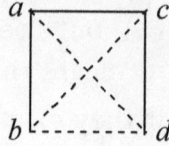

Fig. 2.2.1. Graph $S_{(4)}$.

Theorem on total stratification. Graph G is totally stratified if and only if its subgraphs contain no graph $S_{(4)}$. (A proof is given in Batchelder, Lefebvre, 1982; see also Lefebvre, 2001.)

Let us look at the graphs in Figures 2.2.2 and 2.2.3.

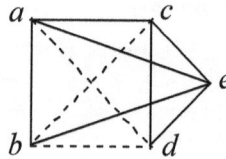

Fig. 2.2.2. A stratified graph that is not totally stratified.

The graph in Fig. 2.2.2 is stratified because it can be represented as $<a,b,c,d>R<e>$. But this graph contains the subgraph $<a,b,c,d>$, which is $S_{(4)}$. Therefore graph $<a,b,c,d,e>$ is not totally stratified.

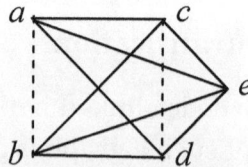

Fig. 2.2.3. A totally stratified graph

None of the four-node subgraphs of the graph in Fig.2.2.3 is

$S_{(4)}$. So, it follows from the theorem on total stratification that the graph in Fig.2.2.3 is totally stratified. Therefore, to determine whether a graph is totally stratified, one needs to find out if there is subgraph $S_{(4)}$ among its subgraphs. If there is no $S_{(4)}$, the graph is totally stratified; if there is a subgraph $S_{(4)}$, it is not. Note that graphs with two or three nodes are stratified.

Consider a graph that is not totally stratified. If we begin taking away its nodes one by one, together with their adjacent edges, in a certain number of steps we will reach a totally stratified graph, since a three-node graph is always totally stratified.

Chapter 3

Decomposable graphs, polynomials, and diagonal forms

In this chapter we present a formal apparatus for analysis of subjects and groups.

3.1. Theorem on decomposition.

The proofs of the following statements are given in Appendix 10 Lefebvre, 2001.

Statement 3.1.1. A graph cannot be stratifiable both in R and in $\overline{R}$ at the same time.

Statement 3.1.2. If a graph is stratifiable in R, its division into minimal strata in R is unique to within strata's numeration.

These statements underlie the procedure for decomposition of a stratifiable graph; we call it the *D*-procedure.

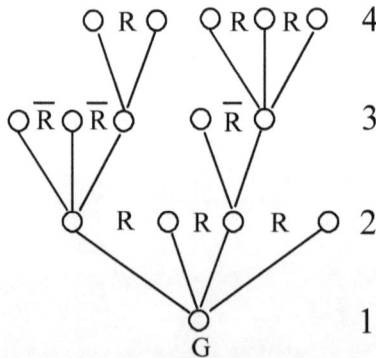

Fig. 3.1.1. Decomposition tree for graph *G*;
tiers numbers are given on the right.

It consists of progressive division of a graph into minimal strata. Each minimal stratum belongs to a certain numbered level of division of tiers. We investigate each minimal stratum that is in relation R with other minimal strata to check whether it is stratifiable in relation $\overline{R}$. (According to the definition of a minimal stratum, it is not stratifiable in R.) If it is not stratifiable in $\overline{R}$, the investigation ends; if it is stratifiable in $\overline{R}$, we divide it into minimal strata in $\overline{R}$, which strata belong to the next tier. This procedure generates trees of the type given in Fig. 3.1.1. Relations R and $\overline{R}$ alternate: R - on the second tier, $\overline{R}$ - on the third, R - on the fourth, etc.

Every circle in Fig.3.1.1 corresponds to a subgraph of graph G; symbols R and $\overline{R}$ depict relations between minimal strata. If a circle is an *end* with no originating branches, it corresponds either to an elementary graph of one node or to a non-stratifiable graph. By Statement 3.1.2, the decomposition tree of a stratifiable graph is unique to within the order of branches originating in circles.

We call a graph *decomposable*, if it is not elementary and every ending circle in its decomposition tree is an elementary graph.

Theorem of decomposition. Graph G is decomposable if and only if it is totally stratifiable. (Batchelder, Lefebvre, 1982; see also Lefebvre, 2001).

It follows from this statement that a graph is decomposable if and only if there is no $S_{(4)}$ among its subgraphs (by the theorem of total stratification).

3.2. Graphs and grammatical trees

Decomposable graphs can be represented in analytical notation to facilitate working with them. Here is the procedure for translating a decomposable graph into analytical notation. First, we construct a decomposition tree. Then, we construct a *grammatical tree*, isomorphic to the decomposition tree. Symbols R and $\overline{R}$ appear in the same places in the grammatical tree as in the decomposition tree. Ends of the branches in the grammatical tree which do not have originating branches are denoted by the letters that correspond to the graph's nodes. Ramifications are denoted by other letters. Letters at the ends of branches are called *end* letters, and others are called *intermediate* letters. Each intermediate letter designates a group of symbols (letters, signs R and $\overline{R}$, parentheses) located immediately above it and can be replaced by this group taken in parentheses. The end result is a *word*, taken to be the analytical equivalent of the graph.

Consider graph in Fig. 3.2.1.

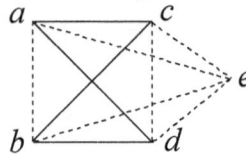

Fig. 3.2.1. Decomposable graph

Solid lines in Fig. 3.2.1. correspond to relation R, and dotted ones to $\overline{R}$. This graph is decomposable because it does not contain the subgraph $S_{(4)}$. Its decomposition tree is given in Fig. 3.2.2, and a grammatical tree for the same graph in Fig. 3.2.3.

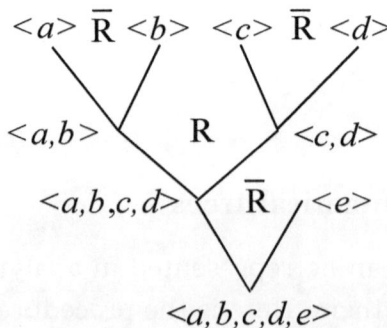

Fig. 3.2.2. Decomposition tree of the graph in Fig. 3.2.1

Letters a, b, c, d and e are end letters, and A_1, A_2, A_3 and A_4 intermediate. A_3 designates expression $a\overline{R}b$, A_4 designates $c\overline{R}d$. We put these expressions in parentheses and substitute them for A_3 and A_4.

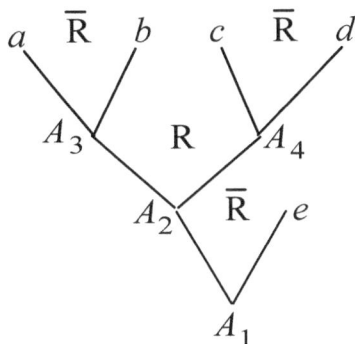

Рис. 3.2.3. Grammatical tree of the graph given in Fig. 3.2.1

Letter A_2 designates $(a\bar{R}b)R(c\bar{R}d)$. We put this expression in parentheses and substitute it for A_2. Finally, we put the expression $((a\bar{R}b)R(c\bar{R}d))\bar{R}e$ in parentheses to replace A_1. The resulting grammatical tree appears as shown in Fig. 3.2.4.

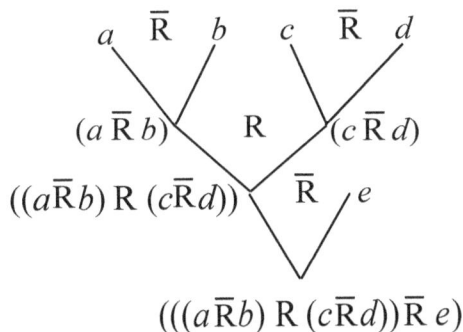

Fig. 3.2.4. Grammatical tree after replacement of the letters
A_1, A_2, A_3 and A_4 by the expressions they designate;
analytical notation at the bottom corresponds to the graph in Fig. 3.2.1

Any decomposable graph can be represented by a similar tree. Each node in the tree corresponds to the analytical notation of some subgraph.

3.3. Polynomials and diagonal forms

Next, we will consider letters in the analytical notation of a graph as variables defined on the set M of all subsets of a universal set and symbols R and $\overline{R}$ as the operations of intersection ($\cdot$) and union (+), defined on the same set. Further, we will substitute $\cdot$ for one of the symbols, R or $\overline{R}$, and + for the other. We will call $\cdot$ *multiplication* and + *addition*. The analytical notation of a graph with the above interpretation is called *polynomial*. A polynomial consisting of one letter is called *elementary* and corresponds to an elementary graph. After substitution $\cdot$ for R and + for $\overline{R}$, the analytical notation in the bottom of Fig. 3.2.4 becomes the following polynomial:

$$(((a + b) \cdot (c + d)) + e). \tag{3.3.1}$$

Polynomials can be written within brackets. Let us agree that instead of $[(A)]$ one can write $[A]$ and instead of $(A \cdot B)$ write $A \cdot B$ or AB. Then, polynomial (3.3.1) appears as follow:

$$[(a + b) (c + d) + e], \tag{3.3.2}$$

and the grammatical tree in Fig. 3.2.4 appears as a tree of polynomials (Fig.3.3.1):

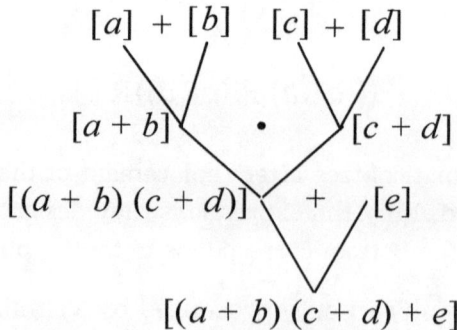

$[a] + [b] \quad [c] + [d]$

$[a + b] \qquad \cdot \qquad [c + d]$

$[(a + b) (c + d)] \qquad + \qquad [e]$

$[(a + b) (c + d) + e]$

Fig. 3.3.1. Tree of polynomials

Now let us eliminate the lines. For each non-elementary polynomial, we will write the polynomials located directly above it, and to the right. Thus, we obtain a treelike object called a *diagonal form*. For the tree in Fig. 3.3.1, the diagonal form appears as follows:

$$
\begin{array}{cccc}
 & [a] + [b] & & [c] + [d] \\
 & [a + b] & & [c + d] \\
 & [(a + b)\,(c + d)] & & & + [e] \\
[(a + b)\,(c + d) + e] & & & &
\end{array}
$$

Expression [a], where a is an elementary polynomial, will be called an elementary diagonal form. *Non-elementary diagonal form is interpreted an exponential formula (see Chapter 1), where the exponent* P^W *corresponds to the function* $\Phi = P + \overline{W}$. *Parentheses and brackets are considered equivalent in computation.*

For the practical purpose of finding a polynomial corresponding to a simple decomposable tree, one may omit the step of explicitly creating a grammatical tree. Also, outside of diagonal forms brackets are not needed.

Consider a graph of three nodes (Fig.3.3.2):

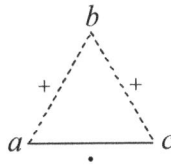

Fig. 3.3.2. A decomposable graph

This graph can be divided into two subgraphs. The first one, consisting of one node, b, is an elementary graph; the second one consists of the two nodes a and c. The node of the first subgraph, b, is in the same relation + with nodes a and c of the second subgraph. This means that the first and second subgraphs are in relation +. We cannot decompose these two subgraphs to smaller subgraphs in relation $+$; thus, they are minimal strata in this relation. The second subgraph can be decomposed to two subgraphs consisting of the nodes a and c in relation $\cdot$. The resulting polynomial is

$$b + ac.$$

Fig. 3.3.3 shows all graphs of the three nodes a, b, c and their corresponding polynomials:

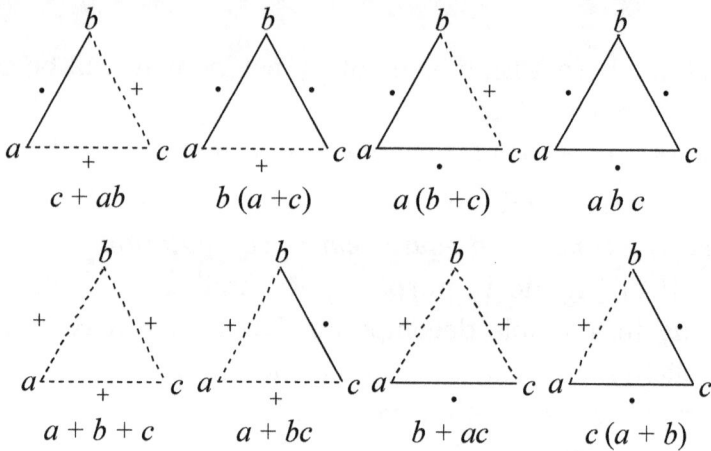

$$c + ab \qquad b\,(a + c) \qquad a\,(b + c) \qquad a\,b\,c$$

$$a + b + c \qquad a + bc \qquad b + ac \qquad c\,(a + b)$$

Fig. 3.3.3. Graphs of three nodes and their corresponding polynomials

Polynomials for decomposable graphs with a greater number of nodes can be found in a similar way. Consider, for example, a decomposable graph of four nodes (Fig.3.3.4):

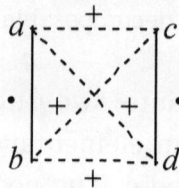

Fig. 3.3.4. A decomposable graph

This graph decomposes into two subgraphs with nodes a, b and c, d. The nodes of the first subgraph are connected with the nodes of the second subgraph by $+$; thus, the subgraphs are in the same relation. Neither subgraph decomposes to smaller subgraph in relation $+$; that is, they are minimal strata in $+$. But each subgraph can be decomposed into elementary subgraphs in relation $\cdot$. Therefore, the graph in Fig. 3.3.4 corresponds to the polynomial

$$ab + cd.$$

Note that a polynomial must correspond to the graph's analytical notation, and we must not perform any transformation of the polynomial, except the commutative, before a diagonal form is constructed.

Let us describe the procedure for construction of diagonal form when a polynomial is known. We will call a polynomial minimal in $\cdot$ if it cannot be represented as a product $A \cdot B$, and minimal in $+$ if it cannot be represented as a sum $A+B$.

1. Enclose the polynomial into brackets.

2. If it is elementary, the procedure ends.

3. If the polynomial can be represented as a product, divide it into multiplier-polynomials minimal in $\cdot$ and write them above and to the right one after another, each enclosed in square brackets.

4. If the polynomial can be represented as a sum, divide it into summand-polynomials minimal in $+$, and write them above and to the right, each in square brackets and connected by $+$.

5. For every polynomial written above and to the right, repeat the procedure starting from 2.

6. The procedure ends when every non-elementary polynomial has diagonal elements.

Consider the graph in Fig.3.3.2. It corresponds to polynomial $b + ac$. Let us construct its diagonal form. First, put the polynomial in brackets:

$$[b + ac].$$

The initial polynomial is a sum of two polynomials minimal in +: b and ac. Put them into brackets, connected by +, and write them above and to the right of the initial polynomial:

$$[b] + [ac]$$
$$[b + ac]$$

.

Polynomial $[b]$ is elementary and cannot be decomposed further, but polynomial $[ac]$ is not elementary, it is a product of two polynomials, a and c. Put them into brackets and write one after another above and to the right of the polynomial $[ac]$:

$$[a]\,[c]$$
$$[b] + [ac]$$
$$[b + ac]$$

. (3.3.3)

The construction of the diagonal form is complete.

Each diagonal form corresponds to a function composed of Boolean operations +, $\cdot$, $^{-}$, and can be represented as

$$Aa + B\overline{a} \qquad (3.3.4)$$

on any of its arguments a, where A and B do not depend on a.

Further on we will use two interpretations of diagonal form. On the one hand, it is a description of the subject with the inner domain; on the other, it is an exponential formula representing a function for computation. Therefore, a diagonal form plays the role of a picture and of a formula.

Chapter 4
Initial model

In this chapter we introduce the basic model that will be used in subsequent analysis. Additional extensions of the model are introduced in Chapter 6.

4.1. A general schema

We assume that a *subject* is an individual or an organization that includes people. Several interacting subjects constitute a group. Every two subjects in a group are either in a relation of cooperation or in one of conflict. Within the framework of our model, the concepts of cooperation and conflict are fundamental and cannot be reduced to other concepts. There is a set of actions that subjects can perform. In the basic model, every subject in the group is able to perform each of those actions.

A subset of actions may or may not be capable of being performed simultaneously. A set of actions is called an *alternative*. A subject chooses an alternative and then realizes any compatible set of actions from the chosen alternative. A subject's choice depends on the relationships among group members and the influences of other subjects. In addition, the subject has an *intention* to choose one or another of the alternatives. The intention is considered as *self-influence*. Subjects are either intentional or not intentional. The former have only intentions that can become reality; the latter have arbitrary intentions.

A subject is capable of integrating the influences of all subjects, including self-influence, into the influence of the group as a whole. For every subject, there is a corresponding diagonal form based on the graph of relationships within the group. The diagonal form describes the subject's reflexion, i.e., the hierarchy of the subject's images of the self, and at the same time represents the function of the subject's choice. A formal procedure for computing the value of this function is regarded as a model of mental choice generation. Relations within the group and influences of other members impose constraints on the subject's choice. The model allows us to predict the subject's possible choices with account taken of these restrictions.

4.2. Representation of the subject

We assume that the subject can perform actions α_1, α_2, . . . , α_S, $S \geq 1$. All of these actions are *acceptable* for the subject, i.e., the subject is able to perform them both technically and morally. *The relation of preference on the set of actions is not given.* The set of actions is regarded as the universal set. Set M of all subsets of the universal set, including the empty set, is the set of alternatives. In other words, each alternative is a subset of the universal set of actions. Choice of the empty set is interpreted as the subject's refusal to choose any non-empty alternative. The subject's activity consists of choosing an alternative from set M, and then realizing the choice. Therefore, the model distinguishes between 'choice' and 'realization of choice'. On the set of all subsets of actions, there is given a unary relation of realization. The empty set and a one-element set can always be realized. For other sets the possibility or impossibility of realization must be specially given. If a non-empty alternative is chosen, then any of its non-empty realizable subsets can be realized (but only one). To clarify the distinction between 'choice' and 'realization', consider an example. Let the universal set consist of two actions:

α_1 – turn left
α_2 – turn right

The Boolean lattice of alternatives appears as follows:

$$1 = \{\alpha_1, \alpha_2\}$$

$$\{\alpha_1\} \qquad \{\alpha_2\}$$

$$0 = \{\ \}$$

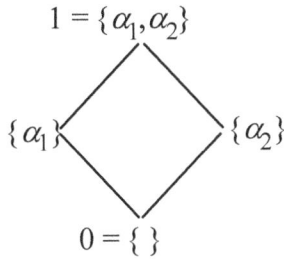

Fig. 4.2.1. Boolean lattice of the set of alternatives

The choice of { } means that the subject chooses inactivity; he will perform neither α_1 nor α_2. The choice of $\{\alpha_1\}$ means that the subject will perform only action α_1, and the choice of $\{\alpha_2\}$ that he will perform only action α_2. Consider the alternative $\{\alpha_1, \alpha_2\}$. Since the subject cannot perform actions, α_1 (turn left) and α_2 (turn right) at the same time, such actions are *incompatible*, and alternative $\{\alpha_1, \alpha_2\}$ *is not realizable*. After choosing it, however, the subject can realize either subset $\{\alpha_1\}$, or subset $\{\alpha_2\}$. The meaning of choice $\{\alpha_1, \alpha_2\}$ is rejection of the empty alternative { }. This is the reason for assigning a unary relation of realization on set M. There are situations in which actions α_1, and α_2 are compatible. For example, α_1 is buying a watch and α_2 is buying a telephone. A person can buy both a watch and a telephone at the same time. In this case, after choosing $\{\alpha_1, \alpha_2\}$, the subject can realize any one of the three subsets of actions: $\{\alpha_1\}$, $\{\alpha_2\}$, $\{\alpha_1, \alpha_2\}$.

A diagonal form corresponds to a particular subject and defines the subject's choice function. Consider subject a_k:

$$\Phi_k = \Phi(a_1, \ ..., \ a_k, \ ..., \ a_n). \tag{4.2.1}$$

Variables $a_1, \ ..., \ a_k, \ ..., \ a_n$ are defined on set M of all subsets of the universal set of actions. The value of function Φ_k is subject a_k's choice. Variable a_i corresponds to subject a_i. The value of a_i is the alternative that a_i inclines a_k to choose. The value of a_k is subject a_k's intention to choose a certain alternative.

As we mentioned earlier, a subject can be intentional or not intentional. The latter can have any intention, i.e., variable a_k can take on any value from M; the subject's choice is given by function (4.2.1). An intentional subject with given values of a_1, ..., a_{k-1}, a_{k+1}..., a_n, has only intentions that can become a choice. Each intention is a solution of the equation (4.2.2):

$$a_k = \Phi_k (\, a_1, \, ..., \, a_k, \, ..., \, a_n).\qquad(4.2.2)$$

The solution is also interpreted as a possible choice. Equation (4.2.2) may have no solution. This is interpreted as the subject's inability to make an intentional choice.

Consider now the connection between diagonal form and the mental aspect of the subject's activity. A diagonal form is constructed by decomposition of the graph of a group containing the subject. The diagonal form represents the subject with a hierarchy of images of the self. It is a tree at whose ramifications there are polynomials in brackets. These polynomials constitute a partially-ordered set; every polynomial located above and to the right of another polynomial is considered to 'follow' the latter. Each polynomial in the diagonal form has its own diagonal form for which it is the bottom-most element. We say that diagonal form B follows diagonal form A, if B's bottom-most polynomial follows A's bottom-most polynomial. In considering diagonal forms as depictions of the subject, we say that B is A's image of the self, if B follows A. This allows us to construct recursive statements of the type "A_m is A_{m-1}'s image of the self that is A_{m-2}' image of the self that is ... A_1's image of the self." Each statement corresponds to the path along the branches from an end to the root. Thus, the chain of statements is finite. Each diagonal form has a unique set of such chains.

As an example, consider diagonal form (4.2.3):

$$
\begin{array}{c}
[a_2]\,[a_3] \\
[a_1] + [a_2\,a_3] \\
[a_1 + a_2\,a_3]
\end{array}
\qquad(4.2.3)
$$

A partial order of polynomials constituting this diagonal form is given in Fig.4.2.2:

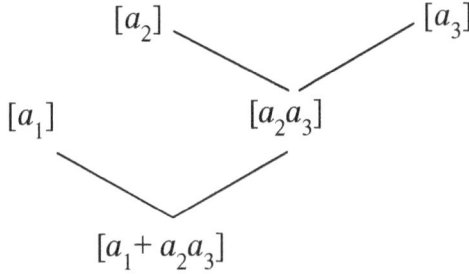

$[a_2]$ $[a_3]$

$[a_1]$ $[a_2 a_3]$

$[a_1 + a_2 a_3]$

Fig. 4.2.2. Partial order of polynomials

A partially-ordered set of diagonal forms is given in Fig. 4.2.3:

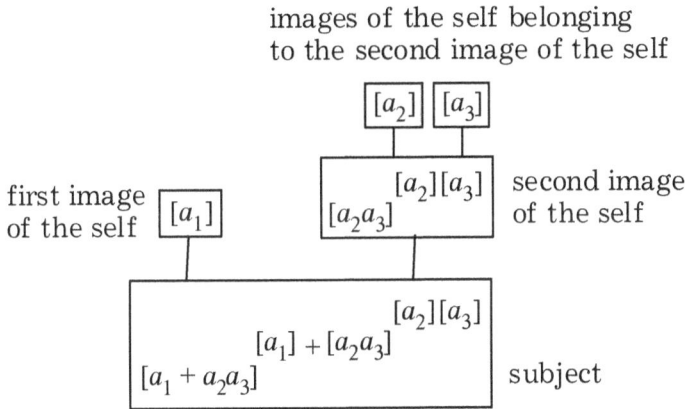

images of the self belonging
to the second image of the self

$[a_2]$ $[a_3]$

first image
of the self $[a_1]$ $[a_2][a_3]$ $[a_2 a_3]$ second image
of the self

$[a_2][a_3]$
$[a_1] + [a_2 a_3]$
$[a_1 + a_2 a_3]$ subject

Fig. 4.2.3. Partial order of diagonal forms

Each lower polynomial corresponds to a group influencing the subject. Symbol · means cooperation, and + confrontation. The value of the polynomial is interpreted as the influence of a group as a whole. The polynomials following the bottom-most one correspond to minimal strata into which the initial group is divided. Each minimal stratum influences the subject's image of the self corresponding to it. Relations between images are predetermined by the relations between minimal strata influencing them. If minimal strata are in a state of cooperation, then the images are in a state of cooperation; if minimal strata are in conflict, then the images are in conflict.

A subject consisting of one letter is called *elementary*. Such a subject corresponds to polynomial [a]; this is also the subject's diagonal form.

A non-elementary subject is depicted by a diagonal form of the type

$$\Phi = P^W, \tag{4.2.4}$$

where P is the bottom-most polynomial of the diagonal form; $W = A_1 * A_2 * ... * A_k$; $k \geq 2$; * either ·, or +, and A_i diagonal forms representing the subject's images of the self.

Expression W is the subject's *integral image of the self* that, in our model, consists of a collection of the images of self in cooperation or conflict with one another. The integral image of the self for subject (4.2.3) is

$$W = [a_1] + [a_2\, a_3]^{[a_2]\,[a_3]}, \tag{4.2.5}$$

where $A_1 = [a_1]$ is the first image of the self; $A_2 = [a_2\, a_3]^{[a_2]\,[a_3]}$ the second image of the self; + means that the images are in conflict.

Expression (4.2.4) gives a function that we call the *reflexion function*. It can be represented as

$$\Phi = P + \overline{W}. \tag{4.2.6}$$

Variables P and W take on values from the set of alternatives. The value of P is interpreted as the alternative to which the group inclines the subject. Actions in set P are called *attractive to the group*, and those in $\overline{P}$ *unattractive to the group*. Let us emphasize that P is the set of actions attractive for the group, whose performance is *expected from a particular subject*. For another subject, the group's influence may be different and the set of expected actions may be different as well. Set W is the result of a choice made by the integral image of the self, that is, the result of the subject's *mental* choice. Since a mental choice of something means preference of this 'something', we say that W consists of actions *attractive for the subject*. Actions in $\overline{W}$ we will call *unattractive for the subject*. The value of Φ corresponds to the subject's choice.

Let us substantiate now our choice of function (4.2.6) as the reflexion function. We will demonstrate that, by means of this function, we insert *the anti-selfishness principle* into the model:

> *While pursuing his own personal goals, the subject may not cause harm to the group he is a member of.*

Within our framework of, this principle can be formulated as follows: it is unacceptable to choose an alternative with actions attractive to the subject but unattractive to the group.

Consider the Venn diagram (Fig.4.2.4).

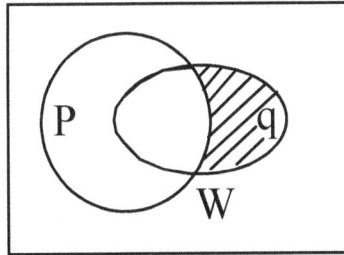

Fig. 4.2.4. Hatched set q is the set of prohibited actions

The set of prohibited actions is as follows:

$$q = \overline{P}W, \tag{4.2.7}$$

where $\overline{P}$ is the set of actions unattractive for the group, and W is the set of actions attractive for the subject.

The set of 'permitted' actions is

$$\overline{q} = P + \overline{W} . \tag{4.2.8}$$

The subject chooses the set of all permitted actions, and, as a result, we obtain the function (4.2.6).

The choices of the subject and of every non-elementary subject in his hierarchy of images are based on the anti- selfishness principle.

Note that an action unattractive to the group may be included in the subject's choice if it is also unattractive to the subject, i.e., if it belongs to the set

$$\overline{P}\,\overline{W} = \overline{P}(P + \overline{W}). \tag{4.2.9}$$

Thus, *the subject may act against the group's interests, if by doing so the subject does not pursue his individual goals and is willing to sacrifice them.*

We have supposed that, in the subject's mental domain, a graph is gradually decomposed to minimal strata and single influences are integrated into the unified influence of the group, represented by the value of the corresponding polynomial. Let set A be the influence of one group, from the subject's point of view, and set B be the influence of another group. (A group may consist of a single individual.) We assume that if these groups are in cooperation, then, from the subject's point of view, they can come to a consensus, so that their joint influence inclines the subject toward choosing actions common to the interests of both groups. Therefore, the influence of groups that are in cooperation corresponds to the *intersection* of sets A and B:

$$AB.$$

If the groups are in conflict, there is no consensus; each group influences the subject independently. Therefore, the influence of groups that are in conflict corresponds to the *union* of the sets A and B :

$$A + B.$$

Similar considerations can be expressed regarding self-images. If images are in cooperation with each other, they are in a consensus; their joint choice is the intersection of their individual choices. If the images are in confrontation, a consensus is impossible; their joint choice is the union of the individual choices.

This substantiates the designation of cooperation by $\cdot$, and conflict by $+$.

4.3. Representation of a group

1. Let us consider a group consisting of subjects a_1, a_2, ..., a_n, где $n \geq 1$.

2. In any non-elementary group, we define the binary relations $\cdot$ (cooperation) and $+$ (conflict) (one of them may be empty). As a result, we obtain the relation graph G. In the framework of the *initial* model, we assume that graph G is decomposable.

3. A set of actions $\{\alpha_1, \alpha_2, ..., \alpha_S\}$, $S \geq 1$, is defined as common to all subjects. This is the universal set 1. A set of all subsets of the universal set (including the empty set) is interpreted as a set of alternatives and designated M.

4. The unary relation of realization is defined on set M for each subject.

5. A matrix of influence is constructed:

$\|p_{ij}\|$, $i =1$, ..., n; $j=1$, ..., n; where $p_{ij} \in M$; p_{ij} is the alternative, which subject a_i inclines subject a_j to choose. An element of the type p_{kk} is subject a_k's self-influence (intention).

6. Using the relation graph, we construct the diagonal form Φ, representing the hierarchy of images of the self and, at the same time, the choice function of each subject in the group.

7. Letters a_1, a_2, ..., a_n are variables. The same function corresponds to each subject:

$$\Phi = \Phi(a_1, a_2, ..., a_n) \, . \qquad (4.3.1)$$

8. If subjects are non-intentional, the value of variables for each of them are predetermined by the matrix $\|p_{ij}\|$, all elements of which are known. Element p_{kk} is interpreted as the subject's intention and, at the same time, as his self-influence. A matrix column number j contains influences on subject a_j. Thus, subject a_j corresponds to expression

$$\Phi_j = \Phi \, (p_{1j}, p_{2j}, ..., p_{nj}) \, . \qquad (4.3.2)$$

Φ_j is an element of M; we interpret it as the alternative chosen by subject a_j. If Φ_j is the empty set, the subject realizes it, i.e., does not perform any actions. If Φ_j is not empty, then the subject realizes any of its non-empty realizable subset, and only one.

9. If subjects are intentional, the diagonal elements, p_{kk}, of their influence matrix $\|p_{ij}\|$ are not known in advance. These elements correspond to the intentions of the subjects constituting a group. They can be found from the following equations for p_{kk}:

$$p_{kk} = \Phi \, (p_{1k}, p_{2k}, ..., p_{kk}, ..., p_{nk}) \, , \qquad (4.3.3)$$

where $k = 1, 2, ..., n$. The value of p_{kk} is interpreted as subject a_k's intention, his self-influence, and his choice, at the same time.

Equation of the type (4.3.3) may or may not have solutions. The absence of solutions means that, with a given relation graph and combination of influences, subject a_k cannot make an intentional choice. In this case, we say that subject a_k is *frustrated* or *in a state of frustration*. It is possible that any element of M is a solution of the equation (4.3.3). Then we say that subject a_k has *freedom of choice* or is *in the state of free choice*. We will say that the subject who can choose only the empty alternative { } is *in a passive state*, and the one who can choose a non-empty alternative is *in an active state*. Sometimes it is convenient to speak of the subject's choice, sometimes of the subject's state, especially in cases where the subject chooses between alternatives 1 and 0.

4.4. Examples for analysis

A group consists of three subjects: a_1, a_2, a_3. Two binary relations, $\cdot$ (cooperation) and $+$ (conflict), are defined. Let subjects (a_1, a_3) and subjects (a_2, a_3) be connected by cooperation, and subjects (a_1, a_2) by conflict. Construct a graph corresponding to this group (Fig.4.4.1):

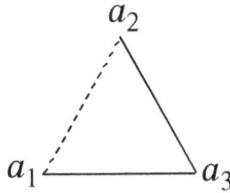

Fig. 4.4.1. Solid lines designate cooperation, and broken lines conflict

Let each subject be able to perform actions α_1, α_2, α_3. The set of actions $\{\alpha_1, \alpha_2, \alpha_3\}$ is the universal set 1. The set of all subsets of the set of actions is M. It is the set of alternatives. A unary relation of realization is given on it. Suppose that this is the same for all subjects: actions α_1 and α_3 cannot be performed at the same time. Thus,

$1 = \{\alpha_1, \alpha_2, \alpha_3\}$ is not realizable
$\{\alpha_1, \alpha_2\}$ is realizable
$\{\alpha_1, \alpha_3\}$ is not realizable
$\{\alpha_2, \alpha_3\}$ is realizable
$\{\alpha_1\}$ is realizable
$\{\alpha_2\}$ is realizable
$\{\alpha_3\}$ is realizable
$0 = \{\,\}$ is realizable

First consider subjects who are non-intentional and construct their own influence matrix. Let subject a_1 have the intention to choose alternative $\{\alpha_2, \alpha_3\}$ and incline subject a_2 to choose 0, and subject a_3 to choose $\{\alpha_2, \alpha_3\}$. Subject a_2 has the intention to choose $\{\,\}= 0$ and inclines a_1 to choose $\{\alpha_1, \alpha_3\}$ and a_3 to choose $\{\alpha_2\}$. Subject a_3 has the intention to choose 1 and inclines a_1 to choose $\{\alpha_2\}$ and a_2 to choose $\{\alpha_1, \alpha_2\}$. Column a_1 contains influences on subject a_1 from the self and from subjects a_2, a_3; column a_2 contains influences on a_2 from the self and a_1, a_3; column a_3 contains influences on a_3 from the self and a_1, a_2.

Table 4.4.1

Matrix of influences in a group of three non-intentional subjects. Lines show influences that subjects exert on others and the self; columns contain influences that are exerted on the subject.

	a_1	a_2	a_3
a_1	$a_1 = \{\alpha_2, \alpha_3\}$	$a_1 = 0$	$a_1 = \{\alpha_2, \alpha_3\}$
a_2	$a_2 = \{\alpha_1, \alpha_3\}$	$a_2 = 0$	$a_2 = \{\alpha_2\}$
a_3	$a_3 = \{\alpha_2\}$	$a_3 = \{\alpha_1, \alpha_2\}$	$a_3 = 1$

The polynomial corresponding to graph in Fig. 4.4.1 is

$$a_3(a_1 + a_2). \tag{4.4.1}$$

Using the procedure of constructing a diagonal form when a polynomial is known, we obtain (4.4.2):

$$\begin{array}{c} [a_1] + [a_2] \\ [a_3]\,[a_1 + a_2] \\ X = [a_3\,(a_1 + a_2)] \end{array} \tag{4.4.2}$$

On the one hand, this form describes the hierarchy of images of the self, the same for all three subjects; on the other, it represents a function of choice, also the same for all three subjects. In the latter case, the diagonal form is regarded at as an exponential formula. It can be simplified:

$$X = a_1 + a_2 + \bar{a}_3 . \tag{4.4.3}$$

The alternatives chosen by subjects a_1, a_2, a_3 will be designated as $X_{a_1}, X_{a_2}, X_{a_3}$, respectively. By substituting values from the first, second, and third columns of the influence matrix into equation (4.4.3), we obtain

$$X_{a_1} = \{\alpha_2, \alpha_3\} + \{\alpha_1, \alpha_3\} + \overline{\{\alpha_2\}} = \{\alpha_1, \alpha_2, \alpha_3\} = 1 ,$$

$$X_{a_2} = 0 + 0 + \overline{\{\alpha_1, \alpha_2\}} = \{\alpha_3\} ,$$

$$X_{a_3} = \{\alpha_2, \alpha_3\} + \{\alpha_2\} + \overline{1} = \{\alpha_2, \alpha_3\} .$$

Subject a_1 chooses set $1 = \{\alpha_1, \alpha_2, \alpha_3\}$ and can realize any non-empty subset of this set of actions except $\{\alpha_1, \alpha_2, \alpha_3\}$ and $\{\alpha_1, \alpha_3\}$, which are non-realizable. Subject a_2 chooses set $\{\alpha_3\}$ consisting of one action α_3. A set of one action is always realizable. Subject a_3 chooses set $\{\alpha_2, \alpha_3\}$, corresponding to three realizable subsets: $\{\alpha_2, \alpha_3\}$, $\{\alpha_2\}$ and $\{\alpha_3\}$. Subject a_3 can realize any of these. Note that the model does not predict which of the realizable subsets will be realized.

Consider now intentional subjects. In this case, the subjects' intentions are not given in advance. They must be found by solving equations. The influence matrix is as follows:

Table 4.4.2

Matrix of influences in a group of three

	a_1	a_2	a_3
a_1	a_1	$a_1 = 0$	$a_1 = \{\alpha_2, \alpha_3\}$
a_2	$a_2 = \{\alpha_1, \alpha_3\}$	a_2	$a_2 = \{\alpha_2\}$
a_3	$a_3 = \{\alpha_2\}$	$a_3 = \{\alpha_1, \alpha_2\}$	a_3

When subjects are intentional, the diagonal elements, intentions a_1, a_2, a_3, are unknown values, unlike the defined sets in the case of non-intentional subjects. The following equations correspond to subjects a_1, a_2, a_3:

$$a_1 = a_1 + a_2 + \bar{a}_3, \tag{4.4.4}$$
$$a_2 = a_1 + a_2 + \bar{a}_3, \tag{4.4.5}$$
$$a_3 = a_1 + a_2 + \bar{a}_3. \tag{4.4.6}$$

These equations result from the successive substitutions of unknown a_1, a_2, a_3 for X in (4.4.3). Transform them to form $a_i = Aa_i + B\bar{a}_i$, $(i = 1, 2, 3)$:

$$a_1 = a_1 + (a_2 + \bar{a}_3)\bar{a}_1, \tag{4.4.7}$$
$$a_2 = a_2 + (a_1 + \bar{a}_3)\bar{a}_2, \tag{4.4.8}$$
$$a_3 = (a_1 + a_2)a_3 + \bar{a}_3. \tag{4.4.9}$$

In the first equation, the unknown is a_1. The values for a_2 and a_3 are taken from the a_1-column of the influence matrix. In the second equation, the unknown is a_2. The values for a_1 and a_3 are taken from the a_2-column. In the third equation, a_3 is unknown. The values of a_1 and a_2 are taken from the a_3-column.

The equation for subject a_1 is

$$a_1 = a_1 + \{\alpha_1, \alpha_3\}\bar{a}_1, \tag{4.4.10}$$

$A = 1$, $B = \{\alpha_1, \alpha_3\}$; $A \supset B$, thus, (4.4.10) is solvable. Its solutions satisfy the inequalities

$$1 \supseteq a_1 \supseteq \{\alpha_1, \alpha_3\}.$$

This implies the existence of two solutions, each of which contains the elements α_1 and α_3:

$$a_1 = \{\alpha_1, \alpha_2, \alpha_3\} = 1,$$
$$a_1 = \{\alpha_1, \alpha_3\}.$$

Subject a_1 can choose any of these solutions and realize any non-empty realizable subset of the chosen set of actions.

The equation for subject a_2 is

$$a_2 = a_2 + \{\alpha_3\}\bar{a}_2, \tag{4.4.11}$$

$A = 1$, $B = \{\alpha_3\}$, $A \supset B$, thus, (4.4.11) has solutions satisfying the inequalities

$$1 \supseteq a_2 \supseteq \{\alpha_3\}.$$

There are four solutions corresponding to the above inequalities, and each of them contains the element α_3.

$$a_2 = \{\alpha_1, \alpha_2, \alpha_3\} = 1,$$
$$a_2 = \{\alpha_1, \alpha_3\},$$
$$a_2 = \{\alpha_2, \alpha_3\},$$
$$a_2 = \{\alpha_3\}.$$

The subject can choose any of the four above sets and then realize any realizable subset of the chosen set.

The equation for the subject a_3 is

$$a_3 = \{\alpha_2, \alpha_3\}a_3 + \bar{a}_3. \tag{4.4.12}$$

$A = \{\alpha_2, \alpha_3\}$, $B = 1$; $A \subset B$, which implies that equation (4.4.12) does not have a solution. Subject a_3 cannot make an intentional choice, and we conclude that he is in a state of frustration.

4.5. The anti-selfishness principle and mathematical formalism

From the formal point of view, the anti-selfishness principle is represented by the following equation:

$$\Phi = P + \overline{W},$$

where Φ is the subject's choice. We will consider below the mathematical formalism corresponding to the mental process of generating P and W.

A non-intentional subject a_k is represented as

$$\Phi(a_1, a_2, ..., a_k, ... a_n) = P + \overline{W} , \qquad (4.5.1)$$

where P, W are functions

$$P = P\,(a_1, a_2, ..., a_k, ..., a_n), \qquad (4.5.2)$$

$$W = W\,(a_1, a_2, ..., a_k, ..., a_n); \qquad (4.5.3)$$

P is the set of actions attractive to the group for subject a_k, and W is the set of actions attractive for subject a_k (see (4.2.6)). By giving the values of influences on subject a_k from other subjects and from the self, we predetermine sets P and W.

For a intentional subject, sets P and W exist only for such collections of variable values a_1, a_2, ..., a_k, ..., a_n, for which the following equation holds:

$$a_k = \Phi(a_1, a_2, ..., a_k, ..., a_n). \qquad (4.5.4)$$

In this case P and W can be considered as functions defined on the set of such collections. We see that the formalism itself generates sets P and W; they are not introduced into the model from without. We do not need to know in advance which actions are attractive to the group and which to the subject.

With fixed values of

$$a_1, a_2, ..., a_{k-1}, a_{k+1}, ..., a_n ,$$

the values of P and W depend on choosing a_k's value that satisfies equation (4.5.4). Thus, the group's preference depends on the subject's choice. This is possible, because it is not a real group, but an image of the group formed in the subject's inner domain. The subject's choice influences his image of the group's preferences. In this way, the subject forecasts actions that would be attractive to the group subsequently to his choice. The subject's own preferences may also depend on his own choice.

To demonstrate how the anti-selfishness principle works in the formal model, we consider a intentional subject a_1 with diagonal form (4.4.2) and equation (4.4.7). Its solutions belong to the interval

$$1 \supseteq a_1 \supseteq (a_2 + \bar{a}_3) . \tag{4.5.5}$$

With values of a_1, a_2 and a_3 satisfying (4.5.5), the set of actions attractive to the group is

$$P = [a_3 (a_1 + a_2)], \tag{4.5.6}$$

and the set of actions attractive to the subject is

$$W = [a_3] [a_1 + a_2] \quad \frac{[a_1] + [a_2]}{} = a_3. \tag{4.5.7}$$

Let the universal set of subject a_1 be $1 = \{\alpha, \beta, \gamma, \delta\}$, and let $a_2 = \{\alpha, \beta\}$ and $a_3 = \{\alpha, \beta, \gamma\}$. Suppose a_1 chooses the alternative $\{\alpha, \beta, \delta\} = a_2 + \bar{a}_3$, from the interval

$$1 \supseteq a_1 \supseteq (a_2 + \bar{a}_3).$$

In this case

$$a_1 = \{\alpha, \beta, \delta\},$$
$$P = \{\alpha, \beta\},$$
$$W = \{\alpha, \beta, \gamma\}.$$

The set of actions prohibited for the subject is

$$\overline{P}W = \{\gamma\} .$$

The set of actions unattractive to the subject is

$$\overline{W} = \{\delta\} .$$

The set of actions unattractive both to the subject and to the group is

$$\overline{PW} = \{\delta\}.$$

The set of actions attractive to the subject and to the group is

$$PW = \{\alpha, \beta\}.$$

We see that set $\{\alpha, \beta, \delta\}$, chosen by the subject, consists of two actions, α and β, attractive both to the subject and to the group, plus one action, δ, unattractive to both the subject and the group. This choice is in accordance with the anti-selfishness principle, since one of the actions included in the choice that is unattractive to the group is also unattractive to the subject.

4.6. Actions that are always chosen and actions that are never chosen.

Let subject a correspond to the equation

$$a = Aa + B\overline{a} , \tag{4.6.1}$$

where $A \supseteq B$. The solutions of this equation belong to the interval

$$A \supseteq a \supseteq B . \tag{4.6.2}$$

If $A = B$, there is only one solution. If $A \supset B$, there are several solutions. In this case, the model does not allow us to predict which one of them will be chosen by the subject. But the model helps us to single out two special sets: the set of actions that are present in any choice, and the set of actions that are absent from all choices. We designate them as R and S, respectively.

The following correlations hold for any solution a satisfying (4.6.2):

$$Ra = R, \tag{4.6.3}$$

$$Sa = 0. \tag{4.6.4}$$

Let us prove first that $R = B$. Any action from B is included in every choice by a belonging to the interval (4.6.2); not a single action from $\overline{B}$ possesses this quality. Thus, B is the set of actions that are present in any choice.

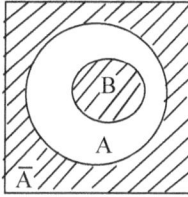

Fig. 4.6.1. Venn diagram. The hatched area in the center (B) is set R; the hatched area outside the larger circle $(\overline{A})$ is set S

We will prove now that $S = \overline{A}$. Not a single action from $\overline{A}$ is included in any choice by the subject, and any action from set A is included at least in one choice. Thus, $\overline{A}$ is the set of actions that are absent from all choices. The Venn diagram clarifies these considerations (Fig. 4.6.1).

Consider an example. Let subject a correspond to the equation

$$a = (b + c)a + c\overline{a}, \tag{4.6.5}$$

the universal set is $\{\alpha, \beta, \gamma, \delta\}$, $b = \{\alpha, \beta\}$ и $c = \{\delta\}$. For these values, we obtain

$$A = \{\alpha, \beta, \delta\}, B = \{\delta\}. \tag{4.6.6}$$

Solutions for equation (4.6.5) belong to the interval

$$\{\alpha, \beta, \delta\} \supseteq a \supseteq \{\delta\}. \tag{4.6.7}$$

It follows from (4.6.7) that equation (4.6.5) has four solution, i.e., the subject has four choices:

$$\{\alpha, \beta, \delta\}, \{\alpha, \delta\}, \{\beta, \delta\}, \{\delta\}.$$

The set of actions always chosen is

$$R = B = \{\delta\},$$

consisting of one action δ. The set of actions never chosen is

$$S = \overline{A} = \overline{\{\alpha,\beta,\delta\}} = \{\gamma\},$$

consisting of one action γ.

In the following, we will consider only intentional subjects.

Chapter 5
Theorems on variety

Let the universal set consist of one action α_1, (i.e., the universal set is $\{\alpha_1\}$); the set of its subsets, M, consists then of two sets: $1=\{\alpha_1\}$ and $0=\{\,\}$. If, under given conditions, the subject can choose only set 1, he is in a state of *activity*; if the subject can choose only set 0, he is in a state of *passivity*; if the subject can choose either 1, or 0, he is in a state of *free choice*; finally, if the subject cannot make a choice, he is in a state of *frustration*. Intuition suggests that there must exist groups, containing subjects in all four states, and groups in which there is a subject capable of being in each of the states, depending on the influences exerted by other subjects. We call these intuitive considerations the *minimal requirements for variety*, which must be satisfied by the model.

5.1. First theorem on variety

Formulation: There is at least one group, with a decomposable relation graph and an influence matrix, such that the group contains subjects in all four states.

Proof. Consider a group of subjects with relation table 5.1.1 and influence table 5.1.2:

Table 5.1.1

Table of relations

	a	b	c	d	e
a		·	+	·	·
b	·		+	·	·
c	+	+		·	·
d	·	·	·		+
e	·	·	·	+	

Table 5.1.2
Table of influences

	a	b	c	d	e
a	a	0	1	0	0
b	1	b	1	0	0
c	0	0	c	1	1
d	0	0	0	d	1
e	0	0	0	0	e

We construct a graph corresponding to the relation table:

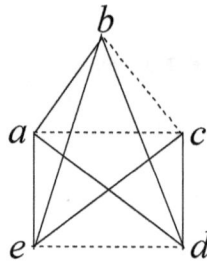

Fig. 5.1.1. Graph of relations

This graph is decomposable. Its polynomial is

$$(ab + c)\,(e + d) \qquad\qquad (5.1.1)$$

and the diagonal form is

$$
\begin{array}{cc}
[a]\,[b] & \\
\quad[ab] \qquad + [c] & [e] + [d] \\
\quad[ab + c] & [e + d] \\
[(ab + c)\,(e + d)] &
\end{array}
$$

After simplification we obtain the equation

$$x = e + d + \overline{c}(\overline{a} + \overline{b}), \qquad\qquad (5.1.2)$$

where x can be replaced by the variables d, e, c and a.
Using Table 5.1.2, find equations for subjects d, e, c and a:

$$d = 0 + d + \overline{1}(\overline{0} + \overline{0}) = d,\tag{5.1.3}$$

$$e = e + 1 + \overline{1}(\overline{0} + \overline{0}) = 1,\tag{5.1.4}$$

$$c = 0 + 0 + \overline{c}(\overline{1} + \overline{1}) = 0,\tag{5.1.5}$$

$$a = 0 + 0 + \overline{0}(\overline{a} + \overline{1}) = \overline{a}.\tag{5.1.6}$$

It follows from these equations that d is in a state of free choice, e is in the active state, c is in the passive state, and a is in a state of frustration. □

5.2. Second theorem on variety

Formulation: There is a group of subjects with a decomposable graph, such that the group contains a subject capable of being in any one of the four states, depending on the influences exerted by other subjects.

Proof. Let the relation table be as shown in Table 5.2.1, and the combination of influences on subject a by b, c, d, e, f be as shown in Table 5.2.2.

Table 5.2.1
Table of relations

	a	b	c	d	e	f
a		+	+	+	.	.
b	+		.	.	.	.
c	+	.		+	.	.
d	+	.	+		.	.
e	.	.	.	.		+
f	.	.	.	.	+	

Table 5.2.2
Four sets of influences

	1	2	3	4
b	1	1	1	0
c	0	1	1	1
d	0	0	1	1
e	1	1	0	0
f	0	1	0	0

Using Table 5.2.1, construct the relation graph (Fig. 5.2.1).

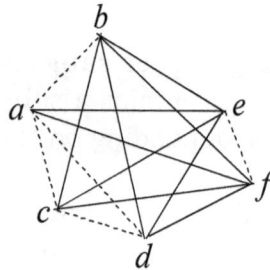

Fig. 5.2.1. Graph of relations

This graph corresponds to the following polynomial:

$$(a + b \, (c + d)) \, (e + f) \qquad (5.2.1)$$

and to the following diagonal form:

$$
\begin{array}{ll}
 & [c] + [d] \\
 & [b]\,[c + d] \\
[a] + [b\,(c + d)] & [e]+[f] \\
[a + b\,(c + d)] & [e+f] \\
[(a + b\,(c + d))\,(e + f)]
\end{array}
\qquad (5.2.2)
$$

We obtain the equation

$$a = (a + b(c + d))(e + f) + \overline{a}\overline{b} \ . \qquad\qquad (5.2.3)$$

By substituting the values of variables from the columns of Table 5.2.2 into equation (5.2.3), we obtain:

$$a = a, \quad a = 1, \quad a = 0, \quad a = \overline{a}.$$

These equalities prove the theorem. $\qquad\qquad\qquad\qquad$ □

Chapter 6
Extension of the initial model

Within the framework of the initial model, we assume that the relations graph is decomposable and that the set of actions is the same for all subjects. In the extended model these limitations are removed. We generalize the model by assuming that the graph of relations may be different for each subject. Finally, we introduce the concept of being *unaware*. To include this concept in the model, we must suppose that variables may have different values on the first and subsequent tiers of the diagonal form.

6.1. Non-decomposable graph of relations

Since the subjects in a group may have arbitrary relations, the graph for a group containing more than three subjects may be non-decomposable. We suppose that, in this case, the subject excludes other subjects from consideration one by one until the graph of relations becomes decomposable. This moment will necessarily arrive, because for three subjects a graph is always decomposable. We assume that each subject has a strict ordering of the other subjects' importance. First, the subject removes the member of the group that is the least important for him. If after that the graph becomes decomposable, the procedure ends. If the graph is still not decomposable, the least important member out of those who left is removed, and so on until the graph becomes decomposable. Note that different subjects may have different orders of importance.

Consider the graph in Fig.6.1.1. This graph is not decomposable, because its subgraph $<a, b, c, d>$ is $S_{(4)}$. Let the importance of other subjects for subject a decrease in the following order:

$$c, b, d, e, \qquad (6.1.1)$$

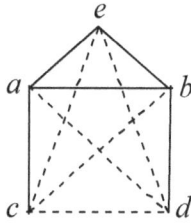

Fig. 6.1.1. Non-decomposable graph

i.e., the most important for a is subject c, and the least important is subject e. The process of removal begins with subject e, after which the graph appears as follows (Fig.6.1.2):

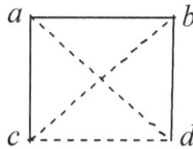

Fig. 6.1.2. Non-decomposable graph

This graph is $S_{(4)}$, so, it is not decomposable. Among the subjects that remain, the least important for a is d. The graph that appears after d's removal contains three nodes (Fig.6.1.3):

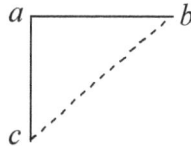

Fig. 6.1.3. Decomposable graph

This graph is decomposable. It corresponds to the polynomial

$$a(b + c).$$

For subject c, let the importance of other subjects decrease in this order:

$$e, a, d, b. \qquad\qquad (6.1.2)$$

Removal begins with subject b, after which we obtain graph (6.1.4):

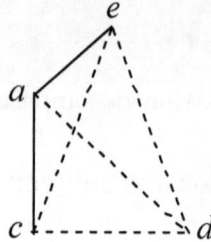

Fig. 6.1.4. Decomposable graph

This graph is decomposable, because there is no $S_{(4)}$ among its subgraphs. It corresponds to the polynomial

$$d + a(e + c).$$

We see that for the list in (6.1.1), subject a needs two removals to transform a non-decomposable graph into a decomposable one. But subject c with list (6.1.2) needs only one removal.

6.2. Individual sets of actions

Every subject may have his own individual set of actions with given realization relations. Consider a group of three subjects: a, b and c.

　　Subject a. Let this subject be able to perform only one action, α_1. The set of his alternatives is

$$1 = \{\alpha_1\},$$
$$0 = \{\ \}.$$

　Subject b. This subject can perform three actions: α_2, α_3, α_4. The set of alternatives is:

$$1 = \{\alpha_2,\ \alpha_3,\ \alpha_4\}$$
$$\{\alpha_2,\ \alpha_3\}$$
$$\{\alpha_3,\ \alpha_4\}$$
$$\{\alpha_2,\ \alpha_4\}$$
$$\{\alpha_2\}$$
$$\{\alpha_3\}$$
$$\{\alpha_4\}$$
$$0 = \{\ \}$$

Subject c. Let c be able to perform two actions, α_5 and $\alpha_6.$ The set of alternatives is

$$1 = \{\alpha_5,\ \alpha_6\},$$
$$\{\alpha_5\},$$
$$\{\alpha_6\},$$
$$0 = \{\ \}.$$

This example demonstrates that different subjects may have different sets of alternatives. In modeling choices in such groups, the influences of group members on each subject must relate to an alternative from the particular subject's set of actions.

6.3. Individual graphs of relations

Until this point, we have looked at relation graphs from an external observer's point of view. Let us analyze a more general scheme and assume that the subjects may have different perceptions of the groups content and of relations among its members.

Consider a set of three subjects . Let each of them differently perceive this collection as a group, as in the example in Fig. 6.3.1. Subjects a and c see all members of the group but have different opinions concerning relations. Subject b does not include a in the group and perceives his relation with c as cooperation, even though c believes it is conflict. In this case, the diagonal forms for each subject constructed differently.

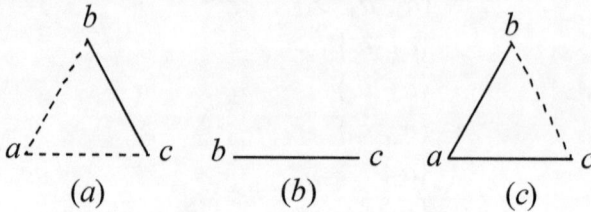

Fig. 6.3.1. Graphs of relations, from the point of view of each subject

6.4. Influences that the subject is aware of and not aware of

Consider a's diagonal form:

$$[a] \, [b]$$
$$[a \, b] \qquad . \qquad (6.4.1)$$

The expressions in brackets on the second tier represent two images of the self that the subject has and is aware of. The expression in brackets on the first tier belongs to a domain that the subject is not aware of. The subject becomes aware of the values of variables a and b on the first tier through their appearance on the second tier. Let us take the next step. Suppose that the influences of other subjects on the first and subsequent tiers may differ. So, in the subject's diagonal form, two variables will correspond to each of the other subjects' influences: one is located on the first tier, and the other on all subsequent tiers. But the intention is the same on all tiers. With this extension, the subject corresponds to the following equation:

$$[a] \, [b_2]$$
$$a = [a \, b_1] \qquad . \qquad (6.4.2)$$

For example, on the second tier, the one that the subject is aware of, a receives b's direction to choose alternative 1, but on the first tier, the one that the subject is not aware of, a receives direction to choose 0, thus

$$a = \overset{[a]\ [1]}{[a\ 0]} \tag{6.4.3}$$

or

$$a = \overline{a} \tag{6.4.4}$$

This means that, as a result of b's influence, subject a will be in a state of frustration.

Chapter 7
Superactivity

The subject will be called *superactive*, if, for all sets of influences from other subjects, the subject chooses alternative 1, i.e., the set containing all actions. A group will be called superactive if every member of the group is superactive.

7.1. Superactive subjects

From the formal point of view, the subject is superactive, if equation

$$a_k = \Phi(a_1, \ldots, a_k, \ldots, a_n) \qquad (7.1.1)$$

corresponding to the subject, has only one solution $a_k=1$ for any set of values of the variables $a_1, \ldots, a_{k-1}, a_{k+1}, \ldots, a_n$. This definition is equivalent to that condition that, for any set of values $a_1, \ldots, a_k, \ldots, a_n$, the following identity holds:

$$\Phi(a_1, \ldots, a_k, \ldots, a_n) \equiv 1 . \qquad (7.1.2)$$

Let us prove this statement. It is clear that (7.1.1) follows from (7.1.2). We will show now that (7.1.2) follows from (7.1.1).

Function $\Phi(a_1, \ldots, a_k, \ldots, a_n)$ can be written as

$$\Phi = A\,(a_1, \ldots, a_{k-1}, a_{k+1}, \ldots, a_n)\,a_k + B(a_1, \ldots, a_{k-1}, a_{k+1}, \ldots, a_n)\,\bar{a}_k ,$$

that is,

$$a_k = Aa_k + B\bar{a}_k . \qquad (7.1.3)$$

Since equation (7.1.1) has only one solution $a_k=1$,

$$A = 1,\ B = 1 \qquad (7.1.4)$$

for any values of variables a_1, ..., a_{k-1}, a_{k+1}, ..., a_n.
Thus, function Φ can be represented as

$$\Phi = a_k + \bar{a}_k \equiv 1,$$

where a_k takes up any value. It follows from this that (7.1.2) is true.

7.2. Superactive groups

Consider the relations graphs in Fig. 7.2.1:

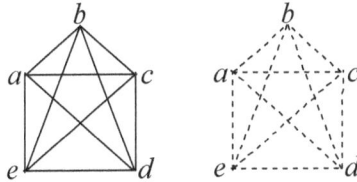

Fig. 7.2.1. Relation graphs of homogeneous groups

The left-hand graph depicts a group in which every two subjects are in a relation of cooperation, and the right-hand graph depicts a group in which every two subjects are in conflict. We will call such groups *homogeneous*. Graphs of homogeneous groups are decomposable; they correspond to polynomials of the type

$$a_1 a_1 \ldots a_n$$

and

$$a_1 + a_1 + \ldots + a_n,$$

Their two-tier diagonal forms are

$$[a_1] [a_2] \ldots [a_n]$$
$$\Phi = [a_1 \ a_2 \ldots \ a_n] \tag{7.2.1}$$

and

$$\Phi = [a_1 + a_2 + ... + a_n] \quad \overline{[a_1] + [a_2] + ... + [a_n]} \tag{7.2.2}$$

or

$$a_1 a_2 ... a_n + \overline{a_1 a_2 ... a_n} \equiv 1, \tag{7.2.3}$$

and

$$a_1 + a_2 + ... + a_n + \overline{a_1 + a_2 + ... + a_n} \equiv 1, \tag{7.2.4}$$

respectively. Therefore, every subject of a homogeneous group is superactive, always chooses alternative 1 and cannot choose any other alternative.

The concept of superactivity allows us to understand the behavior of a *crowd*, considered as a homogeneous group, if we suppose that relations between its member are the same. A crowd's behavior is uncontrollable from within, that is, no subject inside the crowd can influence the behavior of surrounding subjects. This characteristic is described by (7.2.3) and (7.2.4). No subject's choice depends on any other subjects' influences; every subject always chooses alternative 1.

There are other groups, aside from homogeneous groups, that can also be superactive. Consider the graph in Fig. 7.2.2:

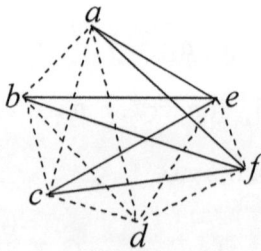

Fig. 7.2.2. Relation graph

This graph is decomposable. It corresponds to the polynomial

$$d + (a + b + c)(e + f) \tag{7.2.5}$$

and the diagonal form

$$[a] + [b] + [c] \qquad [e] + [f]$$
$$[a + b + c] \qquad [e + f]$$
$$[d] + [(a + b + c)(e + f)]$$
$$[d + (a + b + c)(e + f)] \tag{7.2.6}$$

which can be represented as

$$d + (a + b + c)(e + f) + \overline{d + (a + b + c)(e + f)} \equiv 1 .$$

Thus, the group is superactive, although it is not homogeneous.

If the graph of relations is unchanged for long time, then, independently of the subjects' changing influences upon one another, superactivity is a state of long duration.

7.3. Theorem on the impossibility of superpassivity

Let us pose a question: Is there a subject who chooses 0 under all sets of influences from other subjects? Such a subject could be called *superpassive*. For such a subject to exist, an equation of the type (7.1.1) must have the solution $a_k = 0$ for any set of values of the variables $a_1, \ldots, a_{k-1}, a_{k+1}, \ldots, a_n$.

Theorem on the impossibility of superpassivity. For any $n \geq 2$ there is at least one set of values of the variables $a_1, \ldots, a_{k-1}, a_{k+1}, \ldots, a_n$, for which an equation of the type (7.1.1) has the solution $a_k = 1$.

Proof. Function Φ corresponding to any non-elementary diagonal form can be represented as

$$\Phi = P(a_1, \ldots, a_k, \ldots, a_n) + \overline{W}(a_1, \ldots, a_k, \ldots, a_n) . \tag{7.3.1}$$

Function P is a polynomial with operations $\cdot$ and $+$, but without the unary operation of negation. If every variable in (7.3.1) is equal to 1, then $P = 1$, from which

$$\Phi(1, \ldots, a_k=1, \ldots, 1) = 1,$$

i.e., an equation of the type (7.1.1) has solution $a_k=1$ at least for one set of the variables' values.

Therefore, no superpassive subjects can exist. $\qquad\square$

Chapter 8
The paradox of the peacemaker

There is a phenomenon that politicians often observe but do not like to mention: attempts to reconcile conflicting groups often lead to intensification of strain rather than alleviation of the conflict. We will show how our model explains this phenomenon.

8.1. Specifying the concept of influence

Prior to examining in detail the phenomenon mentioned above, we need to define the concept of *influence*, which is broadly used in the framework of our theoretical model. One subject persuades or pressures the other to do something or provokes the other in some way. Or, by some peculiarity, one subject unintentionally inclines the other to perform certain actions; for example, by being weak, he provokes a military attack, or, *vice versa*, by seeming to be strong, provokes retreat without even the need of threats. Thus, subjects may influence each other in arbitrary ways. In the subject's mental domain, the influence of two subjects in a state of cooperation corresponds to the intersection of their individual influences, and the influence of those in conflict to the union of their influences. In other words, we assume that the subject's "mind" perceives the influence of "two friends" as the result of consensus, and the influence of "two adversaries" as the absence of consensus, independently of whether these influences are intentional or not.

In speaking about peacemakers, we will assume that they may incline conflicting parties either to refrain from violence or to use it. The latter does not mean that peacemakers deliberately induce subjects to violent actions; rather, the methods suggested by peacemakers to stop violence may paradoxically lead to more violence. For example, the advice to withdraw troops from a certain region may prompt the residents of that region to rise up in rebellion against the occupiers. We will assume that peacemakers are in relation of cooperation with each member of the conflicting groups, and that among themselves they may be either in a relation of cooperation or in one of conflict.

In our schema, each subject in a conflicting group is in a relation of cooperation with every other subject of the same group and faces a choice between (1) - using violence and (0) - not using violence against members of the other group. The peacemaker also has two alternatives: (1) - to use incentives or (0) - to use punishment. For example, some groups receive economic support and their leaders are given awards, while other groups' economic support is reduced and their leaders are replaced by political competitors.

Every subject in a conflict inclines other subjects to choose one of the alternatives described above and inclines the peacemaker toward using either incentives or punishment.

Let us emphasize: religious or national conflicts may have existed for centuries. The peacemakers' influence, in these cases, is directed not so much toward replacing "conflict" by "cooperation" as toward halting the violence. We will assume that relations among subjects do not change due to the peacemaker's intervention.

8.2. Conflict between two groups of two subjects each

Let the allied countries c and d be in conflict with the allied countries e and f. The graph in Fig.8.2.1 depicts this situation:

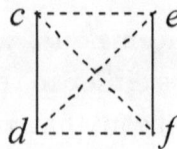

Fig. 8.2.1. Graph of relations in a group of two pairs of allies
in conflict with each other.

This graph is decomposable; it corresponds to the polynomial

$$cd + ef \qquad (8.2.1)$$

and the diagonal form

$$
\begin{array}{cc}
[c]\,[d] & [e]\,[f] \\
[cd] & +\ [ef] \\
[cd + ef] &
\end{array}
\qquad (8.2.2)
$$

Let the subjects choose alternatives from set $M = \{1, 0\}$, where 1 means to use violence and 0 – not to use it. The diagonal form (8.2.2) corresponds to equations of the type $x = cd + ef$, where $x = c, d, e, f$:

$$c = cd + ef,$$

$$d = cd + ef,$$

$$e = cd + ef,$$

$$f = cd + ef.$$

Let the subjects' influences on each other be as follows (Table 8.2.1):

Table 8.2.1

Matrix of influences

	c	d	e	f
c	c	1	0	0
d	1	d	0	0
e	0	0	e	1
f	0	0	1	f

The values of variables are taken from this table: from column c for the first equation, from column d for the second equation, from column e for the third equation, and from column f for the fourth equation. As a result we obtain the equations

$$c = c, \, d = d, \, e = e, \, f = f.$$

We see that subjects c, d, e and f have freedom of choice, i.e., each of them can choose either 1 - (violence) or 0 - (restraint). If all of them choose 0, it would mean that the situation is resolved and the violence stopped without any participation by peacemakers.

Let us investigate what happens if a peacemaker a, in a state of cooperation with all parties, joins the situation (Fig. 8.2.2). The peacemaker inclines each subject either to violence or to refrain from violence. Also, the peacemaker makes decisions: to encourage or to reprimand the various parties to the conflict.

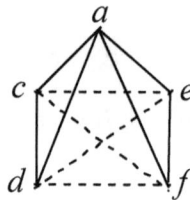

Fig. 8.2.2. Group with a peacemaker

This graph is decomposable; it corresponds to the polynomial

$$a(cd + ef) \tag{8.2.3}$$

and the diagonal form

$$
\begin{array}{c}
[c]\,[d] \quad\ [e]\,[f] \\
[cd] \quad + [ef] \\
[a]\,[cd + ef] \\
[a(cd + ef)]
\end{array}
\tag{8.2.4}
$$

which is transformed into

$$a(cd + ef) + \overline{a(cd + ef)} \equiv 1. \tag{8.2.5}$$

It follows from (8.2.5) that the group consisting of two conflicting pairs becomes superactive following intervention by peacemaker. Subjects c, d, e, f choose 1, i.e., they choose the path of violence.

Therefore, before the peacemaker intervened, violence could have been prevented, but after the intervention the group became superactive, and the path to peaceful resolution was blocked, regardless of any specific influences that were exerted. The peacemaker also chooses 1, i.e., punishment rather than reward.

8.3. Conflict of one subject with a group of two subjects

This situation is depicted by the graph in Fig 8.3.1:

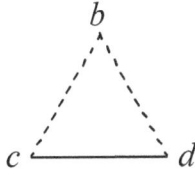

Fig. 8.3.1. Conflict of one subject, b, with group of two subjects, c and d

The corresponding polynomial is

$$b + cd \qquad (8.3.1)$$

and the diagonal form

$$[c]\,[d]$$
$$[b] + [cd]$$
$$[b + cd] \qquad . \qquad (8.3.2)$$

By simplifying (8.3.2), we find equations for subjects b, c and d:

$$b = b + cd,$$
$$c = b + cd, \qquad (8.3.3)$$
$$d = b + cd.$$

The influences are given in Table 8.3.1.

Table 8.3.1
Matrix of influences

	b	c	d
b	b	0	0
c	1	c	0
d	0	0	d

After substituting values from the matrix into equations (8.3.3), we obtain

$$b = b, \; c = 0, \; d = 0.$$

It follows from these equations that b has freedom of choice and can choose either 1 or 0; c and d choose 0. Thus, the group corresponding to Fig.8.3.1 has the potential to refrain from violence.

What will happen if peacemaker a joins the group (Fig.8.3.2)?

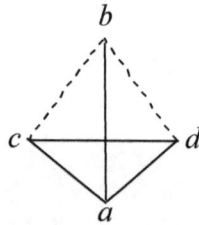

Fig. 8.3.2. Graph of group from Fig. 8.3.1 with peacemaker a

This graph corresponds to the polynomial

$$a \, (b + cd) \tag{8.3.4}$$

and the diagonal form

$$[c]\,[d]$$
$$[b] + [cd]$$
$$[a]\,[b + cd]$$
$$[a\,(b + cd)] \qquad\qquad , \qquad\qquad (8.3.5)$$

which after transformation becomes

$$a(b + cd) + \overline{a(b + cd)} \equiv 1. \qquad (8.3.6)$$

The group whose graph is given in Fig. 8.3.2 is superactive. Subjects b, c, and d choose alternative 1, i.e., violence. Peacemaker a also chooses alternative 1; for him it means using punishment to prevent violence.

Therefore, the inclusion of a peacemaker in a group consisting of one subject in conflict with two subjects, themselves in a relationship of cooperation, makes the group superactive, and the possibility of refraining from violence, previously present, is lost.

8.4. Conflict of two subjects

In the previous two sections, we gave examples of increased tension resulting from the addition of a peacemaker to groups that are in conflict. Nevertheless, there are examples in which the peacemaker's mediation is very useful, like a marriage counselor who helps conflicting spouses.

Consider the conflict between two elementary groups, that is, two subjects (Fig.8.4.1):

$$b \bullet - - - - - - - - \bullet c$$

Fig. 8.4.1. Conflict between two subjects

This graph corresponds to the polynomial

$$b + c \qquad (8.4.1)$$

and the diagonal form

$$\begin{array}{c} [b] + [c] \\ [b + c] \end{array} \qquad , \qquad (8.4.2)$$

which is equivalent to

$$b + c + \overline{b + c} \equiv 1.$$

We see that the group in Fig. 8.4.1 is superactive.

With the addition of a peacemaker, the graph becomes Fig.8.4.2:

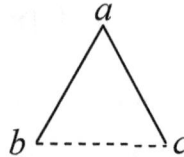

Fig. 8.4.2. Peacemaker a joins conflicting subjects b and c

This graph corresponds to the polynomial

$$a(b + c) \qquad (8.4.4)$$

and the diagonal form

$$\begin{array}{c} [b] + [c] \\ [a]\,[b + c] \\ [a\,(b + c)] \end{array} \qquad , \qquad (8.4.5)$$

which is transformed into

$$a(b + c) + \overline{a} . \qquad (8.4.6)$$

As a result, we obtain three equations:

$$\begin{aligned} b &= a(b + c) + \overline{a}, \\ c &= a(b + c) + \overline{a}, \qquad (8.4.7) \\ a &= a(b + c) + \overline{a}. \end{aligned}$$

Is there now an opportunity to refrain from violence? Consider the following matrix of influences:

Table 8.4.1

Matrix of influences

	a	*b*	*c*
a	*a*	1	1
b	0	*b*	0
c	0	0	*c*

The peacemaker, *a*, inclines subjects *b* and *c* toward violence; they incline each other toward refusal of violence and the peacemaker toward refusal of punishment. With these influences, the equations appear as follows:

$$b = b, \quad c = c, \quad a = \bar{a}.$$

Subjects *b* and *c* have freedom of choice and are able to choose 0, a nonviolent course of action. The peacemaker is in a state of frustration. Therefore, when two subjects are in conflict, the presence of a peacemaker can avert violence.

8.5. Generalization

The analysis conducted in sections 8.2, 8.3 and 8.4 demonstrates that a peacemaker's success depends on the type of group he tries to reconcile. If two subjects, themselves in cooperation, are in conflict with another two subjects who are in a state of cooperation with each other, or if one subject is in conflict with a pair of subjects who are in a state of cooperation, then the presence of a peacemaker may stimulate violence. However, if two elementary groups, that is, two individuals, are in conflict, then the presence of a peacemaker may avert violence. Let us generalize this result.

A group consisting of conflicting subgroups corresponds to

the polynomial

$$(a_1...a_{k_1})+...+(r_1...r_{k_n}),\qquad(8.5.1)$$

where max $(k_1, ..., k_n) > 1$.

Using (8.5.1), we construct the diagonal form

$$[(a_1...a_{k_1})+...+(r_1...r_{k_n})]\quad \begin{matrix}[a_1]...[a_{k_1}]\\ [a_1...a_{k_1}]\end{matrix}\quad +\;.\;.\;.\;.\;+\;\begin{matrix}[r_1]...[r_{k_n}]\\ [r_1...r_{k_n}]\end{matrix}\;.\qquad(8.5.2)$$

After transformation, form (8.5.2) is equivalent to (8.5.1).

An equation for subject a_1 is as follows:

$$a_1 = (a_1...a_{k_1})+...+(r_1...r_{k_n}).\qquad(8.5.3)$$

Let the values of all variables on the right-hand side of (8.5.3), except a_1, be equal to 0. Then (8.5.3) is reduced to one of the following:

$$a_1 = 0,\; a_1 = a_1.\qquad(8.5.4)$$

The second equation appears when $k_1 = 1$. Each equation has the solution 0. Thus, subject a_1 is able to choose the nonviolent alternative. Similar considerations can be made for each subject. Therefore, a group consisting of conflicting subgroups, among which at least one is non-elementary, can transfer into a state where the subjects can make the decision not to use force.

Let us look at what happens if peacemaker z joins the group. The polynomial changes to

$$z((a_1...a_{k_1})+...+(r_1...r_{k_n}))\qquad(8.5.5)$$

and the diagonal form to

$$\frac{[a_1]...[a_{k_1}] \qquad\qquad [r_1]...[r_{k_n}]}{\frac{[a_1...a_{k_1}] + ... + [r_1...r_{k_n}]}{\frac{[z][(a_1...a_{k_1})+...+(r_1...r_{k_n})]}{[z((a_1...a_{k_1})+...+(r_1...r_{k_n}))]}}}} \qquad , \qquad (8.5.6)$$

which after transformation becomes

$$z((a_1...a_{k_1}) + ... + (r_1...r_{k_n})) + \overline{z((a_1...a_{k_1}) + ... + (r_1...r_{k_n}))} \equiv 1 . \qquad (8.5.7)$$

Thus, the group of conflicting subgroups becomes superactive after a peacemaker joins it: each subject chooses violence, and the peacemaker chooses punishment.

Consider a group in which every individual is in conflict with everybody else. This group corresponds to the polynomial

$$a_1 + . . . + a_n \qquad (8.5.8)$$

and the diagonal form

$$\frac{[a_1] + . . . + [a_n]}{[a_1 + . . . + a_n]} \qquad , \qquad (8.5.9)$$

which is transformed into

$$a_1+...+a_n + \overline{a_1+...+a_n} \equiv 1 . \qquad (8.5.10)$$

A group consisting of n conflicting individuals is superactive. After peacemaker z joins the group, its polynomial is

$$z (a_1 + . . . + a_n) , \qquad (8.5.11)$$

and its diagonal form

$$\frac{[a_1] + . . . + [a_n]}{\frac{[z] [a_1 + . . . + a_n]}{[z (a_1 + . . . + a_n)]}} \qquad . \qquad (8.5.12)$$

After simplifying the form we obtain the expression $z(a_1 +...+ a_n) + \bar{z}$ and equations corresponding to the subjects:

$$a_1 = z(a_1 +...+ a_n) + \bar{z} ,$$
$$a_2 = z(a_1 +...+ a_n) + \bar{z} ,$$

. . . .   (8.5.13)

. . . .

$$a_n = z(a_1 +...+ a_n) + \bar{z} ,$$
$$z = z(a_1 +...+ a_n) + \bar{z} .$$

Let the peacemaker provoke subjects a_1, a_2, ..., a_n to choose violence. Let each subject incline every other subject to reject violence (since they are united by hostility toward the peacemaker) and incline the peacemaker to choose encouragement. The matrix of these influences is given in Table 8.5.1.

Table 8.5.1

Matrix of influences

	a_1	a_2	. . . . .	a_n	z
a_1	a_1	0	. . . . .	0	0
a_2	0	a_2	. . . . .	0	0
. . .	. . . .				
a_n	0	0	. . . . .	a_n	0
z	1	1	. . . . .	1	z

By substituting the values from the matrix of influences into equation (8.5.13) (from column a_1 into the equation for a_1, from column a_2 into the equation for a_2 and so on), we obtain solutions for (8.5.13):

$$a_1 = a_1, \ a_2 = a_2, \ . \ . \ . \ . \ , \ a_n = a_n, \ z = \bar{z} . \quad (8.5.14)$$

We see that every subject has acquired freedom of choice and can choose a nonviolent line of behavior. This is possible thanks to the presence of the peacemaker and his provocative influences. The peacemaker himself is in a state of frustration.

Thus, a peacemaker's presence in a group of conflicting subgroups increases the intensity of conflict, but when single individuals are in conflict, the peacemaker's influence can reduce that intensity.

8.6. The case of two peacemakers

Consider a conflict between the two pairs of subjects represented in Fig. 8.2.1. Let two peacemakers a and b, in a state of cooperation, join this group (Fig.8.6.1).

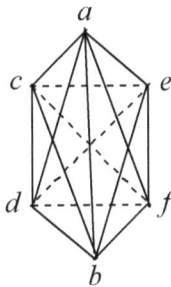

Fig. 8.6.1. Group containing a pair of subjects in conflict
and two peacemakers

This graph is decomposable. It corresponds to the polynomial

$$ab\ (cd + ef) \qquad\qquad (8.6.1)$$

and the diagonal form

$$
\begin{array}{cc}
[c]\,[d] & [e]\,[f] \\
[cd] & +\ [ef] \\
\end{array}
$$
$$[a]\,[b]\,[cd\ + ef]$$
$$[ab\ (cd + ef)] \qquad\qquad\qquad . \qquad (8.6.2)$$

By simplifying (8.6.2), we find that

$$ab(cd + ef) + \overline{ab(cd + ef)} \equiv 1 . \tag{8.6.3}$$

The group in Fig.8.6.1 is superactive, and the paradox of the peacemaker appears. The addition of a second peacemaker does not eliminate the paradox.

Consider now a situation in which peacemakers a and b are in the state of confrontation between themselves (Fig.8.6.2).

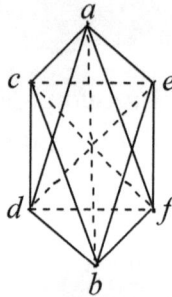

Fig. 8.6.2. Group containing two conflicting subgroups
and two conflicting peacemakers

This graph is decomposable; it corresponds to the polynomial

$$(a + b)\,(cd + ef), \tag{8.6.4}$$

and the diagonal form

$$
\begin{array}{cccc}
 & & [c]\,[d] & [e]\,[f] \\
 & [a] + [b] & [cd] & + [ef] \\
 [a + b] & & [cd + ef] & \\
 [(a + b)\,(cd + ef)] & & &
\end{array}
\tag{8.6.5}
$$

which is transformed into

$$(a + b)(cd + ef) + \overline{(cd + ef)} = a + b + (\overline{c} + \overline{d})(\overline{e} + \overline{f}) . \tag{8.6.6}$$

As a result, we obtain six equations of the type

$$x = a + b + (\bar{c} + \bar{d})(\bar{e} + \bar{f}),\qquad(8.6.7)$$

where x can be replaced with the variables a, b, c, d, e, f.

Let peacemakers a and b incline all subjects to nonviolent action and each other to encouragement. Subjects c, d, e, f incline each other toward violence and the peacemakers to punishment. These influences are shown in Table 8.6.1:

Table 8.6.1

Matrix of influences

	a	b	c	d	e	f
a	a	0	0	0	0	0
b	0	b	0	0	0	0
c	1	1	c	1	1	1
d	1	1	1	d	1	1
e	1	1	1	1	e	1
f	1	1	1	1	1	f

By substituting $x = a$, $x = b$, $x = c$, $x = d$, $x = e$, $x = f$ and the corresponding values of variables from the matrix of influences into equation (8.6.7), we find that

$$a = a, \ b = b, \ c = 0, \ d = 0, \ e = 0, \ f = 0.$$

We see that the peacemakers have freedom of choice; they can choose either punishment or encouragement; each of the subjects c, d, e, f chooses nonviolence. Therefore, a group consisting of two pairs of allies being in conflict *can be* inclined to choose nonviolent action, if the two peacemakers joining the group are in conflict with each other.

Chapter 9

Reflexive Control

Reflexive control is a way to influence subjects that inclines them to make decisions predetermined by the controlling party (Lefebvre, 1965, 1967; Taran, Shemaev, 2004). Within the framework of our model, any influence on a subject or a group of subjects can be considered reflexive control. We distinguish four types of reflexive control:

– manipulation by influence;
– manipulation by changing relations;
– manipulating by order of significance (possible only if the group's graph is not decomposable);
– influence over a subject's inner domain without the subject's knowledge.

In this chapter, we will assume that each subject chooses an alternative from set {0, 1}.

9.1. Manipulation by influence

First, we will consider *direct* influence. Here is the form of such influence:

a wants b to choose x and exerts influence x.

Consider the graph of relations (Fig. 9.1.1):

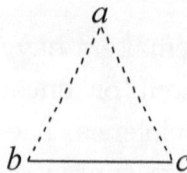

Fig. 9.1.1. Graph of relations

It corresponds to the polynomial

$$a + bc \qquad (9.1.1)$$

and to the diagonal form

$$
\begin{array}{c}
[b]\,[c] \\
[a] + [bc] \\
[a + bc]
\end{array} \qquad (9.1.2)
$$

After transformation, we see that this form is equivalent to the polynomial (9.1.1.). The equation for subject b can be written:

$$b = a + bc \qquad (9.1.3)$$

or

$$b = (a + c)b + a\overline{b}\,; \qquad (9.1.4)$$

$$A = a + c,\ B = a,\ A \supseteq B,$$

hence, all values of b from the interval

$$(a + c) \supseteq b \supseteq a \qquad (9.1.5)$$

are solutions to (9.1.4).
    When $c = 0$,

$$a \supseteq b \supseteq a, \qquad (9.1.6)$$

thus,

$$b = a. \qquad (9.1.7)$$

If a wants b to choose 1, a's influence on b must be 1; if a wants b to choose 0, a's influence on b must be 0.
    When $c = 1$,

$$1 \supseteq b \supseteq a\,. \qquad (9.1.8)$$

If $a = 1$, then $b = 1$, i.e., b obeys a's requirement. If $a = 0$, then b has freedom of choice and will not necessarily obey a. Thus a can incline b to choose 1, but cannot make him choose 0.

Consider reverse influence, when a subject chooses the alternative opposite to the one another subject inclines him. Here is the form of such influence:

> a wants b to choose x, but to achieve this a must exert influence $\bar{x}$.

Let the graph of relations be as follows (Fig.9.1.2):

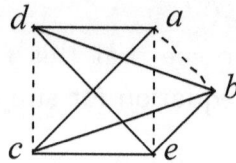

Fig. 9.1.2. Graph of relations

This graph is decomposable; it corresponds to the polynomial (9.1.9)

$$(c + d)\,(a + be), \tag{9.1.9}$$

and the diagonal form (9.1.10),

$$[b]\,[e]$$
$$[c] + [d] \qquad [a] + [be]$$
$$[c + d] \qquad [a + be]$$
$$[(c + d)\,(a + be)] \tag{9.1.10}$$

After transformation, we obtain the equation for b (9.1.11):

$$b = c + d + \bar{a}(\bar{b} + \bar{e}). \tag{9.1.11}$$

If $c = 0$, $d = 0$ and $e = 0$, then b's choice is given by the function

$$b = \bar{a} . \tag{9.1.12}$$

Therefore, within the framework of our model, reverse reflexive control is possible.

Let us raise a question: is it possible to influence a subject such that he moves into the state of free choice? Here is the form of such influence:

> *a* wants *b* to have freedom of choice and exerts
> influence *x*.

Consider the graph in Fig. 9.1.1. Subject *b* corresponds to equation (9.1.3). Let $c = 1$, and let *a* incline *b* to choose 0. As a result we obtain the equation

$$b = b. \qquad (9.1.13)$$

This equation has two solutions: $b = 1$ and $b = 0$. Thus, by inclining *b* to choose 0, *a* moves him into a state of free choice. If *a* inclines *b* to choose 1, *b* chooses 1 and does not have the freedom of choice.

In the above example the controlling subject inclines the controlled subject to choose 0, in order to move him into a state of free choice. We will demonstrate now that there is also a case where the subject moves into a state of free choice after being pushed toward 1. Consider again the graph in Fig.9.1.1. Let *c* exercise reflexive control over *b* instead of *a*. In this case we will also have equation (9.1.3). For $a = 0$,

$$b = bc. \qquad (9.1.14)$$

If the controlling subject *c* pushes *b* toward choosing 1,

$$b = b, \qquad (9.1.15)$$

i.e., subject *b* acquires freedom of choice.

Let us consider the situation of a subject in the state of *frustration* and unable to make a decision. This happens when the equation corresponding to the subject does not have a solution. Here is the form:

> *a* wants *b* to become incapable of making a choice and
> exerts influence *x*.

A group corresponds to the graph in Fig. 5.1.1. We write the equation for b, assuming that $x = b$ in (5.1.2):

$$b = e + d + \overline{c}(\overline{a} + \overline{b}).$$ (9.1.16)

Let $e = 0$, $d = 0$, $c = 0$; then

$$b = \overline{a} + \overline{b}.$$ (9.1.17)

If subject a influences b to choose 1, the equation will be

$$b = \overline{b},$$ (9.1.18)

thus, b is in a state of frustration.

It is possible that subject b moves into a state of frustration after being pushed toward 0. Let the relations graph be as follows:

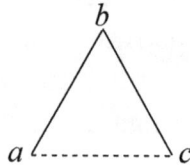

Fig. 9.1.3. Graph of relations

This graph corresponds to the polynomial

$$b\,(a + c),$$ (9.1.19)

and the diagonal form

$$\begin{array}{c} [a] + [c] \\ [b]\,[a + c] \\ [b\,(a + c)] \end{array},$$ (9.1.20)

and the equation for b is

$$b = b\,(a + c) + \overline{b}.$$ (9.1.21)

If $c = 0$,

$$b = ba + \overline{b}.$$ (9.1.22)

To move b into a state of frustration, a must incline b toward choosing 0, which leads to equation (9.1.18).

Consider the case in which a group of two subjects, a and c, controls subject b. Here is the first schema:

a and c want b to choose x; to achieve this is a inclines b to choose y, and c incline b to choose z.

Construct the following graph (Fig. 9.1.4):

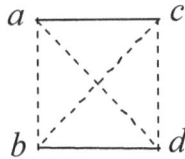

Fig. 9.1.4. Graph of relations

This graph corresponds to the polynomial

$$ac + bd, \qquad (9.1.23)$$

and the diagonal form

$$\begin{array}{cc} [a]\,[c] & [b]\,[d] \\ [ac] & + [bd] \\ [ac + bd] & \end{array} \qquad , \qquad (9.1.24)$$

and the equation for b is

$$b = ac + bd . \qquad (9.1.25)$$

Let $d = 0$; then

$$b = ac. \qquad (9.1.26)$$

We see that, in this example, b's choice is predetermined by a and c. If at least one of them pushes b toward 0, b will choose 0; if both of them, a and c, incline b toward 1, b will choose 1.

Here is the second schema:

a and c want b to have the freedom of choice; to accomplish this, a exerts influence x, and c exerts influence y.

Consider the graph in Fig. 9.1.1. Equation (9.1.3) corresponds to b. To give b the freedom of choice, a must push b toward choosing 0, and c must push b toward 1.

Let us note that, in our model, subjects cannot push other subjects toward states of frustration or freedom of choice. They can only incline a subject toward choosing one or another alternative such that the subject moves into a predetermined state. If a subject is superactive, manipulation by influence is ineffective, since under any set of influences the subject chooses alternative 1.

9.2. Manipulation by changing relations

The main feature of this type of reflexive control is that the graph of relations is modified by the controlling subject. It is important to note that such modification may affect the choices of other subjects as well.

We single out two types of manipulation. In the first type, controlling subject a changes the relation (a, b), in the second, a leaves the group.

Here is the form of this manipulation:

a wants b to obtain freedom of choice and so modifies the relation (a, b).

The initial graph of relations is given in Fig.9.2.1:

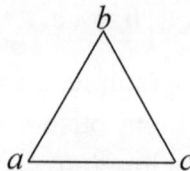

Fig. 9.2.1. Graph of relations

It corresponds to the polynomial

$$a\, b\, c \qquad (9.2.1)$$

and the diagonal form

$$[a\, b\, c] \genfrac{}{}{0pt}{}{[a]\,[b]\,[c]}{} \equiv 1. \qquad (9.2.2)$$

This group is superactive. Thus, b chooses only 1. The graph does not allow b to obtain the freedom of choice due to any modification of influence by other subjects.

What will happen if a changes the relation (a, b) from cooperation to conflict (Fig. 9.2.2)?

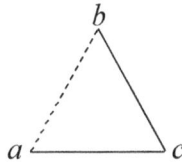

Fig. 9.2.2. Graph of relations after change in relation (a, b)

This new graph corresponds to the polynomial

$$c\,(a + b), \qquad (9.2.3)$$

and the diagonal form

$$[c\,(a + b)] \genfrac{}{}{0pt}{}{[a] + [b]}{[c]\,[a + b]} , \qquad (9.2.4)$$

and the equation for b is

$$b = b + a + \bar{c} . \qquad (9.2.5)$$

For $a = 0$ and $c = 1$, the equation becomes

$$b = b. \qquad (9.2.6)$$

With the given values of a and c, subject b has freedom of choice. Therefore, after the relation has changed, b is able to obtain freedom of choice.

Consider the following schema:

> a wants b to lose the possibility of obtaining freedom of choice and changes the relation (a, b).

Let the initial situation be depicted by the graph in Fig.9.2.2. Subject a changes the relation (a, b) from conflict to cooperation. As a result, the new situation corresponds to the graph in Fig. 9.2.1. Subject b becomes superactive and loses the possibility to have freedom of choice.

We have demonstrated in Chapter 8 that a group corresponding to the graph in Fig.8.2.2 is superactive, i.e., every subject in the group can choose only 1.

Consider the following schema:

> a wants c to be able to have freedom of choice and, to achieve this, a leaves the group.

As a result, the graph of relations changes to the one in Fig. 8.2.1. This graph corresponds to the diagonal form (8.2.2), and the equation for c is

$$c = cd + ef. \qquad (9.2.7)$$

We see that after a leaves the group, c can have freedom of choice, for example, if $d = 1$ and $e = 0$.

9.3. Manipulation by order of subjects' significance

When a graph of relation is not decomposable, the subject ranks other subjects by order of significance and removes them one by one, starting from the least significant, until the graph becomes decomposable. When a manipulates b's ranking of others' significance, there is the danger of becoming so insignificant for b

that b removes a, that is, b would cease to notice a. To avoid this, a has to be careful to remain in the graph.

Consider an example. A graph of relations is as follows (Fig. 9.3.1):

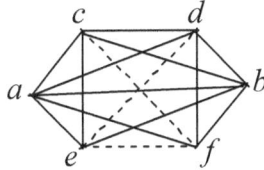

Fig. 9.3.1. Non-decomposable graph

This graph is not decomposable, because it contains subgraph $S_{(4)}$: $<e, c, d, f>$. The order of significance for subject b is $c, d, f, e,$ a. The first subject removed is a. The resulting graph $<b, c, d, e, f>$ is not decomposable either, because it contains the same subgraph $S_{(4)}$: $<e, c, d, f>$. The next subject removed is e, resulting in a new graph that is decomposable (Fig.9.3.2):

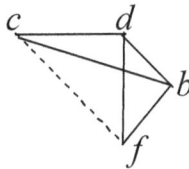

Fig. 9.3.2. Decomposable graph

This graph corresponds to the polynomial

$$b\,d\,(c + f),\qquad(9.3.1)$$

and the diagonal form

$$\begin{array}{c} [c] \; + [f] \\ [b]\,[d]\,[c + f] \\ [b\,d\,(c + f)] \end{array} \qquad (9.3.2)$$

and the equation for b is

$$b = c + f + \bar{b} + \bar{d} \qquad (9.3.3)$$

or

$$b = (c + f + \bar{d})b + \bar{b} . \qquad (9.3.4)$$

Equation (9.3.4) has a solution if

$$(c + f + \bar{d}) \supseteq b \supseteq 1 . \qquad (9.3.5)$$

It follows from (9.3.5) that equation (9.3.4) has a solution only under the condition that $c + f + \bar{d} = 1$. This solution is $b = 1$. When $c + f + \bar{d} = 0$, equation (9.3.4) does not have a solution, and subject b is in a state of frustration.

In (9.3.5), there is no variable corresponding to subject a, thus b's ability to make a choice does not depend on a. But a can avoid being removed.

The schema is as follows:

> a does not want to be removed from b's graph and so changes the order of other subjects' significance for b.

Let a change b's order of significance to the following: c, d, e, a, f. Now, subject a occupies the second to last place and subject f occupies the last place. The least significant for b is f, so, b removes f. As a result, the graph in Fig.9.3.1 changes to the graph in Fig.9.3.3:

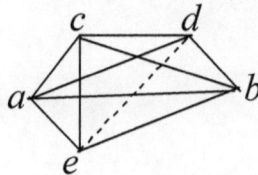

Fig. 9.3.3. Graph of relations after removing subject f

This graph is decomposable. It corresponds to the polynomial

$$a\,b\,c\,(e + d),\qquad(9.3.6)$$

and the diagonal form

$$
\begin{array}{c}
[e] + [d] \\
[a]\,[b]\,[c]\,[e + d] \\
[a\,b\,c\,(e + d)]
\end{array}
,\qquad(9.3.7)
$$

and the equation for b is

$$b = e + d + \overline{a} + \overline{b} + \overline{c}\qquad(9.3.8)$$

or

$$b = (e + d + \overline{a} + \overline{c})b + \overline{b}\,,\qquad(9.3.9)$$

which contains the variable a, such that subject a can influence b's choice. This equation has a solution if

$$e + d + \overline{a} + \overline{c} = 1\,.\qquad(9.3.10)$$

In this case $b = 1$. We see that by inclining b to choose 0, subject a eliminates the case in which b would not be able to make a choice.

9.4. Reflexive control by subconsciousness influence

Consider now the case of unconscious influence (see Chapter 6). There is a group of two subjects a and b in conflict between themselves. Its diagonal form is

$$
\begin{array}{c}
[a] + [b] \\
[a + b]
\end{array}
.\qquad(9.4.1)
$$

We endow b with a subconscious by regarding the letters a on the first and second tiers as independent variables: a_1 is a's unconscious influence on b, and a_2 is the influence of which b is aware. Subject b corresponds to equation

$$b = [a_1 + b] \quad {}^{[a_2 + b]} \qquad \qquad (9.4.2)$$

Here is the schema:

   a wants *b* to have freedom of choice

To achieve this goal, *a* exercises influence on both the conscious and subconscious levels. At the subconscious level, *a* inclines *b* to choose 0; *b* is not aware of this ($a_1 = 0$). At the conscious level, *a* inclines *b* to choose 1, and *b* is aware of it ($a_2 = 1$). Substitute these values into (9.4.2) and we find that

$$b = [0 + b] \quad {}^{[1 + b]} \quad = b , \qquad \qquad (9.4.3)$$

that is, subject *b* has freedom of choice.

Chapter 10

Personal relations

In this chapter, we will show how to analyze relations between individuals with the help of the reflexive games theory.

10.1. Son, mother, father

A son is contemplating marriage. He may marry either α or β. The son has good relations with both his mother and his father, but their relations are tense. The mother inclines him to choose α, and the father wants him to choose β. Can the son make a choice, and, if so, which one? This situation is depicted in the following graph:

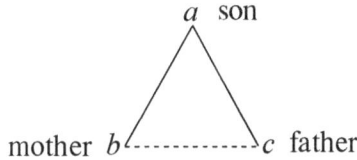

Fig. 10.1.1. Graph of relations

The universal set consists of two actions:

$$\alpha - \text{marry } \alpha ,$$
$$\beta - \text{marry } \beta .$$

Set M consists of four alternatives:

$1 = \{\alpha, \beta\}$ not realizable (we assume a society where polygamy is not practiced)

$\{\alpha\}$ realizable

$\{\beta\}$ realizable

$0 = \{\ \}$ realizable

Alternative 1 means "to marry" without specifying whom; 0 means "not to marry." The following polynomial corresponds to the graph in Fig. 10.1.1:

$$a (b + c),$$ (10.1.1)

the diagonal form

$$[b] + [c]$$
$$[a] [b + c]$$
$$[a (b + c)]$$, (10.1.2)

and the equation for the son is

$$a = (b + c) a + \bar{a}.$$ (10.1.3)

This equation has a solution only if $b + c = 1$. The value of b is the mother's influence; the value of c is the father's influence. In the given case, $b = \{\alpha\}$, $c = \{\beta\}$. Since $\{\alpha\} = \overline{\{\beta\}}$, then $\{\alpha\} + \{\beta\} = 1$, and equation (10.1.3) has a solution

$$a = \{\alpha, \beta\} = 1.$$ (10.1.4)

With the given influences, the son chooses alternative 1, i.e., makes the decision to marry without specifying whom. Set $\{\alpha\}$, as well as set $\{\beta\}$, is realizable.

Let us now see what would happen if both mother and father inclined the son not to marry. Then $b = 0$, $c = 0$. After substituting these values into (10.1.3), we obtain

$$a = \bar{a}.$$ (10.1.5)

This equation has no solution; thus, the son is in a state of frustration and cannot make a choice. The son will also be in frustration if both mother and father strongly incline him to choose the same alternative, for example, α. In this case, $b = \{\alpha\}$, $c = \{\alpha\}$, and equation (10.1.3) is

$$a = \{\alpha\} a + \bar{a},$$ (10.1.6)

where $A = \{\alpha\}$, $B = 1$. Condition $A \supseteq B$ is not met, thus (10.1.6) has no solution.

If father (or mother) pushes the son choose 1={α, β} (to marry), then the son will choose 1, because in this case $A=1$, $B=1$, i.e., equation (10.1.3) has solution the $a = 1$.

Let us change the relations between mother, father and son: the son is now in conflict with both parents, and the parents have good relations. Their graph of relations is as follows:

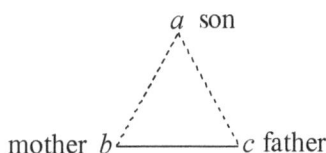

Fig. 10.1.2. Graph of relations

This graph corresponds to the polynomial

$$a + bc, \qquad (10.1.7)$$

and the diagonal form

$$\begin{array}{c} [b]\,[c] \\ [a] + [bc] \\ [a + bc\;] \end{array} \qquad , \qquad (10.1.8)$$

and the equation for the son is

$$a = a + bc\,\bar a\,, \qquad (10.1.9)$$

$A = 1$, $B = bc$, $A \supseteq B$. Equation (10.1.9) has solutions for any values of b and c; therefore, the son cannot be in a state of frustration.

Let mother and father incline the son to choose different alternatives: b ={α}, c ={β}. Since

$$\{\alpha\}\,\{\beta\} = 0, \qquad (10.1.10)$$

we can write (10.1.9) as

$$a = a, \qquad (10.1.11)$$

i.e., the son has freedom of choice.

If mother and father incline the son to choose the same alternative: $b = \{\alpha\}$, $c = \{\alpha\}$, equation (10.1.9) becomes

$$a = a + \{\alpha\}\bar{a} , \qquad (10.1.12)$$

$A = 1$, $B = \{\alpha\}$. Solutions of this equation satisfy the inequalities

$$1 \supseteq a \supseteq \{\alpha\} . \qquad (10.1.13)$$

Hence, the two solutions to equation (11.1.12) are:

$$a = 1 = \{\alpha, \beta\},$$
$$a = \{\alpha\}.$$

Therefore, the model predicts that the son can choose one of two alternatives: either make the decision to marry ($a = \{\alpha, \beta\}$), or choose α ($a = \{\alpha\}$). If he chooses 1, to realize his choice he will need to pick either α or β. If the son chooses alternative $\{\alpha\}$, this means that he decides to marry α and no longer considers β.

Let us analyze now the case where there are three candidates: α, β and γ. The universal set consists of three actions and set M contains eight alternatives:

$$1 = \{\alpha, \beta, \gamma\} \quad \text{not realizable,}$$
$$\{\alpha, \beta\} \quad \text{not realizable,}$$
$$\{\alpha, \gamma\} \quad \text{not realizable,}$$
$$\{\beta, \gamma\} \quad \text{not realizable,}$$
$$\{\alpha\} \quad \text{realizable,}$$
$$\{\beta\} \quad \text{realizable,}$$
$$\{\gamma\} \quad \text{realizable,}$$
$$0 = \{\} \quad \text{realizable.}$$

Consider the scenario related to the graph in Fig.10.1.1. Let the mother incline the son to choose $\{\alpha\}$, and the father $\{\beta\}$. We saw that with such influences in the case of two candidates the son makes the decision to marry. What happens if there are three candidates? Substitute values $b = \{\alpha\}$ and $c = \{\beta\}$ into (10.1.3) and we obtain

$$a = (\{\alpha\} + \{\beta\})a + \bar{a}$$

or

$$a = \{\alpha, \beta\}a + \bar{a}, \tag{10.1.14}$$

where $A = \{\alpha, \beta\}$, $B = 1 = \{\alpha, \beta, \gamma\}$. Thus, relation $A \supseteq B$ does not hold and equation (10.1.14) does not have a solution. This means that the son cannot make a choice and is in a state of frustration. Therefore, the third available action may radically change the subject's ability to make a choice.

Here is another case. The mother does not want the son to marry α, and the father does not want the son to marry β. In the framework of our model, we consider "unwillingness" to choose x as equal to pressure toward $\bar{x}$. Thus, if the mother does not want the son to choose $\{\alpha\}$, she inclines him to choose $\overline{\{\alpha\}} = \{\beta, \gamma\}$, and the father's unwillingness to see the son choose $\{\beta\}$ means that he pushes the son toward $\overline{\{\beta\}} = \{\alpha, \gamma\}$. What will the son choose? The mother's and father's influences are

$$b = \{\beta, \gamma\},$$
$$c = \{\alpha, \gamma\}. \tag{10.1.15}$$

In the case of the scenario in Fig.10.1.1, the substitution of these values results in the equation

$$a = (\{\beta, \gamma\} + \{\alpha, \gamma\})a + \bar{a} \tag{10.1.16}$$

or

$$a = \{\alpha, \beta, \gamma\}a + \bar{a}, \tag{10.1.17}$$

or

$$a = a + \bar{a} = 1. \tag{10.1.18}$$

Therefore, the son makes the decision to marry, without specifying whom.

What will be the son's choice in the case of a scenario related to Fig.10.1.2? Substitution of the values from (10.1.15) into equation (10.1.9) results in

$$a = a + \{\beta, \gamma\}\{\alpha, \gamma\}\bar{a} \qquad (10.1.19)$$

or

$$a = a + \{\gamma\}\bar{a} .$$

This equation has the solution:

$$1 \supseteq a \supseteq \{\gamma\} .$$

All subsets of set $\{\alpha, \beta, \gamma\}$ that contain the element γ are located between 1 and $\{\gamma\}$:

$$1 = \{\alpha, \beta, \gamma\}, \{\alpha, \gamma\}, \{\beta, \gamma\}, \{\gamma\}. \qquad (10.1.20)$$

The son can choose any of these subsets.

Consider now examples of reflexive control over the son's choice where the universal set is $\{\alpha, \beta, \gamma\}$. The mother wants the son to marry either α, or β, and the father does not want the son to marry at all. Their relations are given in Fig.10.1.1. What influence must the father exert over the son for him not to marry? Substitute the value of the mother's influence, $b = \{\alpha, \beta\}$, into equation (10.1.3):

$$a = (\{\alpha, \beta\} + c)a + \bar{a} . \qquad (10.1.21)$$

The father controls the value of c and can use it to exercise reflexive control. He cannot make the son to choose 0, because there is no value of c for which equation (10.1.21) has the solution 0, but the father can move the son into a state of frustration by inclining him to choose one of the following alternatives:

$$\{\alpha, \beta\}, \{\alpha\}, \{\beta\}, \{ \}. \qquad (10.1.22)$$

Since condition $A \supseteq B$, necessary for equation (10.1.21) to have a solution, does not hold if c takes values from set (10.1.22), the son will be in a state of frustration and will not make any choice, including the choice "to marry."

Let us analyze reflexive control aimed at the son's unconscious domain. If the mother has good relationships with father and son, but the father and son are in conflict, their relations are depicted in Fig.10.1.3:

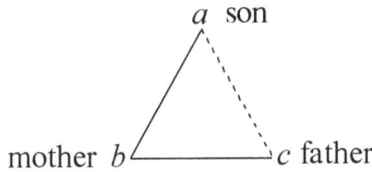

Fig. 10.1.3. Graph of relations

This graph corresponds to the polynomial

$$b\,(a + c), \tag{10.1.23}$$

and the equation for a is

$$a = \dfrac{[a] + [c]}{[b]\,[a + c]}{[b\,(a + c)]} \tag{10.1.24}$$

Let the mother influence the son on both the conscious and unconscious levels (see section 9.4). The mother is represented by letter b. These letters are independent variables on the first and second tiers: b_1 and b_2. Equation (10.1.24) changes to

$$a = \dfrac{[a] + [c]}{[b_2]\,[a + c]}{[b_1\,(a + c)]} \tag{10.1.25}$$

If the mother does not want the son to marry, her goal is to make him to choose 0. To achieve this, the values must be $b_1 = 0$, $b_2 = 1$. Therefore, influencing on the subconscious level the mother has to incline the son to remain unmarried, but on the conscious level she must incline him to marry.

10.2. Escape from jail

Five inmates - John, Tom, Peter, Bob, and Larry - are contemplating an escape. The graph of their relations is given in Fig.10.2.1:

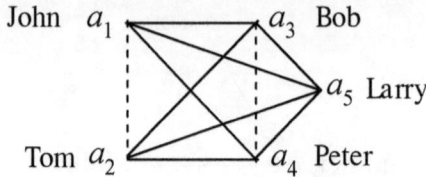

John a_1 a_3 Bob

a_5 Larry

Tom a_2 a_4 Peter

Fig. 10.2.1. Graph of relations

We see that John (a_1) and Tom (a_2) are in conflict, and Bob (a_3) and Peter (a_4) are in conflict. Larry has good relations with everyone. Each inmate – a_1, a_2, a_3, a_4, a_5 – has his own plan of escape. We designate these plans as α_1, α_2, α_3, α_4 и α_5, respectively. Every inmate tries to persuade the others to accept his plan. What will be their individual choices? The graph in Fig.10.2.1 is decomposable. It corresponds to the polynomial

$$a_5 (a_1 + a_2) (a_3 + a_4) \tag{10.2.1}$$

and the diagonal form

$$
\begin{array}{ccc}
& [a_1] + [a_2] & \quad [a_3] + [a_4] \\
[a_5] [a_1 + a_2] & & [a_3 + a_4] \\
[a_5 (a_1 + a_2) (a_3 + a_4)] & &
\end{array}
\tag{10.2.2}
$$

Every plan is an action. Alternatives $\{\alpha_1, \alpha_2, \alpha_3, \alpha_4, \alpha_5\}$ constitute the universal set. The power set of the alternatives, M, consists of the $2^5 = 32$ subsets of this set. We assume that the plans are incompatible with each other. A choice of $1 = \{\alpha_1, \alpha_2, \alpha_3, \alpha_4, \alpha_5\}$ is interpreted as decision "to escape" (independently from the plan), and a choice of $0 = \{ \}$ as a decision to reject an escape. Set M has only six realizable alternatives:

$$\{\alpha_1\}, \{\alpha_2\}, \{\alpha_3\}, \{\alpha_4\}, \{\alpha_5\}, \{ \ \} \ . \tag{10.2.3}$$

The influences are given in the following table:

Table 10.2.1

Matrix of influences

	a_1	a_2	a_3	a_4	a_5
a_1	a_1	$\{\alpha_1\}$	$\{\alpha_1\}$	$\{\alpha_1\}$	$\{\alpha_1\}$
a_2	$\{\alpha_2\}$	a_2	$\{\alpha_2\}$	$\{\alpha_2\}$	$\{\alpha_2\}$
a_3	$\{\alpha_3\}$	$\{\alpha_3\}$	a_3	$\{\alpha_3\}$	$\{\alpha_3\}$
a_4	$\{\alpha_4\}$	$\{\alpha_4\}$	$\{\alpha_4\}$	a_4	$\{\alpha_4\}$
a_5	$\{\alpha_5\}$	$\{\alpha_5\}$	$\{\alpha_5\}$	$\{\alpha_5\}$	a_5

Diagonal form (10.2.2) corresponds to equations of the type

$$x = a_5(a_1 + a_2)(a_3 + a_4) + \overline{a}_5, \tag{10.2.4}$$

where $x = a_1$, a_2, a_3, a_4, a_5. To compute the values of x for each inmate, we take the values of other variables from the corresponding columns. Let us start from Larry, (a_5): $x = a_5$. The values of other variables are taken from the last column of the matrix of influences. As a result, we obtain equation for a_5:

$$a_5 = a_5(\{\alpha_1\} + \{\alpha_2\})(\{\alpha_3\} + \{\alpha_4\}) + \overline{a}_5 \tag{10.2.5}$$

or

$$a_5 = \{\alpha_1, \alpha_2\}\{\alpha_3, \alpha_4\}a_5 + \overline{a}_5 \ . \tag{10.2.6}$$

Since

$$\{\alpha_1, \alpha_2\} \{\alpha_3, \alpha_4\} = 0, \tag{10.2.7}$$

equation (10.2.6) becomes

$$a_5 = \overline{a}_5 . \tag{10.2.8}$$

Thus, Larry is in a state of frustration and cannot make a choice.

Consider now John (a_1); for $x = a_1$, the values of other variables are taken from the first column of the matrix of influences and substituted for variables in equation (10.2.4):

$$a_1 = \{\alpha_5\}(a_1 + \{\alpha_2\})(\{\alpha_3\} + \{\alpha_4\}) + \overline{\{\alpha_5\}} \qquad (10.2.9)$$

or

$$a_1 = \overline{\{\alpha_5\}}, \qquad (10.2.10)$$

since

$$\{\alpha_5\}(\{\alpha_3\} + \{\alpha_4\}) = 0 .$$

Similarly, find equations and solutions for the other inmates

$$a_2 = \overline{\{\alpha_5\}}, \qquad (10.2.11)$$
$$a_3 = \overline{\{\alpha_5\}}, \qquad (10.2.12)$$
$$a_4 = \overline{\{\alpha_5\}}, \qquad (10.2.13)$$

Therefore, four inmates reject Larry's plan and choose the alternative

$$\overline{\{\alpha_5\}} = \{\alpha_1, \alpha_2, \alpha_3, \alpha_4\}, \qquad (10.2.14)$$

then, they try to reach an agreement about which plan to realize. The new universal set is $1 = \{\alpha_1, \alpha_2, \alpha_3, \alpha_4\}$. Set M consists of sixteen alternatives: 1 means the decision "to escape," 0 means not to escape.

Let Larry propose not escaping and let the other inmates keep insisting on their own plans. The new matrix is given in Table 10.2.2. All the rows except the last one contain the same values as in Table 10.2.1. In the last row, there is a new instance of pressure by Larry (a_5). He inclines the others to choose alternative 0, that is, to reject the idea of exodus. The pressure on Larry by other inmates does not change, so that, equations (10.2.5) and (10.2.8) correspond him in the new situation as well, and Larry continues in a state of frustration.

Table 10.2.2

Matrix of influences

	a_1	a_2	a_3	a_4	a_5
a_1	a_1	$\{\alpha_1\}$	$\{\alpha_1\}$	$\{\alpha_1\}$	$\{\alpha_1\}$
a_2	$\{\alpha_2\}$	a_2	$\{\alpha_2\}$	$\{\alpha_2\}$	$\{\alpha_2\}$
a_3	$\{\alpha_3\}$	$\{\alpha_3\}$	a_3	$\{\alpha_3\}$	$\{\alpha_3\}$
a_4	$\{\alpha_4\}$	$\{\alpha_4\}$	$\{\alpha_4\}$	a_4	$\{\alpha_4\}$
a_5	0	0	0	0	a_5

Let us write the equation for John. We substitute values from column a_1 of the matrix of influences (Table 10.2.2) into equation (10.1.4) and obtain

$$a_1 = 0(a_1 + \{\alpha_2\})(\{\alpha_3\} + \{\alpha_4\}) + \overline{0}, \qquad (10.2.15)$$

i.e.,

$$a_1 = 1.$$

Similarly, write equations for the other inmates and find that the choices of Tom (a_2), Bob (a_3), and Peter (a_4) are the same as John's (a_1):

$$a_2 = 1, \qquad (10.2.16)$$

$$a_3 = 1, \qquad (10.2.17)$$

$$a_4 = 1. \qquad (10.2.18)$$

So, Larry is in a state of frustration, and other inmates choose alternative 1, to escape; that is, they again reject the plan suggested by Larry.

Our analysis demonstrates that a group may contain an outcast whose plans are always rejected. The reason is Larry's "total friendliness." Let us test this hypothesis.

We change the scenario and put Larry in conflict with

everyone. The universal set is $1 = \{\alpha_1, \alpha_2, \alpha_3, \alpha_4, \alpha_5\}$. The graph of relations is given in Fig.10.2.2.

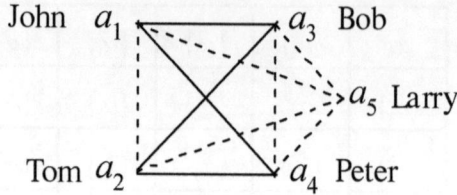

John a_1 a_3 Bob a_5 Larry   Tom a_2 a_4 Peter

Fig. 10.2.2. Graph of relations

This graph is decomposable. It corresponds to the polynomial

$$a_5 + (a_1 + a_2)(a_3 + a_4) \tag{10.2.19}$$

and the diagonal form

$$
\begin{array}{cc}
[a_1] + [a_2] & [a_3] + [a_4] \\
[a_1 + a_2] & [a_3 + a_4]
\end{array}
$$
$$[a_5] + [(a_1 + a_2)(a_3 + a_4)]$$
$$[a_5 + (a_1 + a_2)(a_3 + a_4)] \tag{10.2.20}$$

After transformation of (10.2.20), we obtain

$$a_5 + (a_1 + a_2)(a_3 + a_4) + \overline{a_5 + (a_1 + a_2)(a_3 + a_4)} \equiv 1, \tag{10.2.21}$$

i.e., the group becomes superactive, and each inmate, including Larry, chooses the same alternative

$$1 = \{\alpha_1, \alpha_2, \alpha_3, \alpha_4, \alpha_5\}, \tag{10.2.22}$$

meaning "to escape," regardless of the specific plan.

We see that in the new scenario, where Larry is in conflict with everybody, he is not in a state of frustration, and the others do not reject his plan.

10.3. Theft

Imagine now that only Larry escapes and is not caught. The other four blame each other for not escaping, and their relations change. John (a_1) and Tom (a_2) become allies in conflict with Bob (a_3) and Peter (a_4), who are also allies. The new relations are presented in Fig.10.3.1:

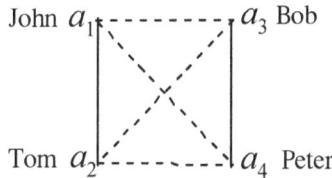

Fig. 10.3.1. Graph of relations

This graph corresponds to the polynomial

$$a_1 a_2 + a_3 a_4, \tag{10.3.1}$$

and the diagonal form

$$[a_1 a_2 + a_3 a_4] \quad [a_1 a_2] \quad [a_1] [a_2] \quad + [a_3 a_4] \quad [a_3] [a_4] \tag{10.3.2}$$

All of a sudden, John (a_1) notices that his well-hidden cigarette has disappeared. He begins to suspect that one of his cellmates stole it. It could be his new friend, Tom (a_2), who knew where the cigarette was hidden. On the other hand, it is natural to suspect that John's enemies, Bob (a_3) or Peter (a_4), may have stolen it. Besides, John admits that he might have lost his cigarette or smoked it and forgot that he did so.

Let John try to figure out what happened to the cigarette. John's universal set contains the following actions:

$$\alpha_2 - \text{to blame Tom,}$$
$$\alpha_3 - \text{to blame Bob}$$
$$\alpha_4 - \text{to blame Peter.}$$

Set M consists of eight alternatives, all of which are realizable:

$1 = \{\alpha_2, \alpha_3, \alpha_4\}$ John suspects Tom, Bob, and Peter,

$\quad\{\alpha_2, \alpha_3\}$ John suspects Tom and Bob

$\quad\{\alpha_2, \alpha_4\}$ John suspects Tom and Peter

$\quad\{\alpha_3, \alpha_4\}$ John suspects Bob and Peter

$\quad\{\alpha_2\}$ John suspects Tom

$\quad\{\alpha_3\}$ John suspects Bob

$\quad\{\alpha_4\}$ John suspects Peter

$0 = \{\ \}$ nobody is suspected; the cigarette was not stolen.

After choosing a non-empty alternative, John can realize any non-empty set of actions contained in that alternative. Let us write the equation for John. It follows from (10.3.2) that

$$a_1 = a_1 a_2 + a_3 a_4. \tag{10.3.3}$$

John suspects everyone in his cell, that is, each of them unknowingly inclines John to suspect himself. Thus,

$$a_2 = \{\alpha_2\},\ a_3 = \{\alpha_3\},\ a_4 = \{\alpha_4\}. \tag{10.3.4}$$

By substituting these values into (10.3.3), we obtain

$$a_1 = a_1 \{\alpha_2\} + \{\alpha_3\} \{\alpha_4\}, \tag{10.3.5}$$

or

$$a_1 = a_1 \{\alpha_2\}. \tag{10.3.6}$$

This equation can be rewritten as

$$a_1 = \{\alpha_2\} a_1 + \{\ \} \bar{a}_1, \tag{10.3.7}$$

where $A = \{\alpha_2\}$ and $B = \{\ \} = 0$. Since $A \supset B$, the equation has the solutions

$$\{\alpha_2\} \supseteq a_1 \supseteq \{\}, \qquad (10.3.8)$$

and John can choose either of the alternatives:

$$\{\alpha_2\},$$
$$\{\} = 0. \qquad (10.3.9)$$

The choice of $\{\alpha_2\}$ means that John would think his friend Tom stole the cigarette; a choice of $\{\}$ means that John would think that the cigarette was not stolen.

10.4. The boss and the award

There are five people working in an office: Alex, Bart, David, Gregory, and Edward, who is their boss. Edward's superior informs him that he can nominate one of his subordinates for a departmental award. Edward's group knows that one of the members may be nominated and that they can influence this decision. The graph of relations in the group is shown in Fig. 10.4.1:

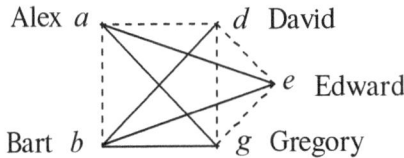

Fig. 10.4.1. Graph of relations

Edward's universal set consists of four actions: α, β, δ and γ, where α means to nominate Alex, β to nominate Bart, δ to nominate David, and γ to nominate Gregory. Alex and Bart do not think of the young and inexperienced David as their competitor, and to demonstrate their disinterest, they pressure Edward to choose David. They are sure that David will not be nominated. Gregory and David will be happy if any person from the office is nominated. So,

$$a = \{\delta\}, \; b = \{\delta\}, \; g = \{\alpha, \beta, \gamma, \delta\} = 1, \; d = \{\alpha, \beta, \gamma, \delta\} = 1. \qquad (10.4.1)$$

The graph in Fig. 10.4.1 is not decomposable, because it contains subgraph $S_{(4)}$: <a, b, g, d>. To make a choice, Edward must remove the least significant coworkers one by one. Let their significance diminish in the following order:

Alex, Bart, Gregory, David.

The least significant is David; Edward removes him, and the graph in Fig.10.4.1 now appears as follows:

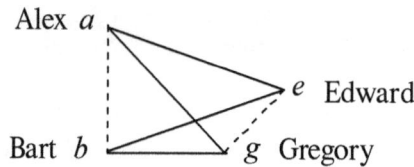

Fig. 10.4.2. Graph of relations after David's removal

This graph is decomposable and corresponds to the polynomial

$$(a + b)(e + g), \tag{10.4.2}$$

and the diagonal form

$$
\begin{array}{cc}
[a] + [b] & [e] + [g] \\
[a + b] & [e + g] \\
[(a + b)(e + g)] &
\end{array}
\tag{10.4.3}
$$

and the equation for Edward is

$$e = (a + b)(e + g). \tag{10.4.4}$$

By substituting the values of variables from (10.4.1) into this equation, we obtain

$$e = (\{\delta\} + \{\delta\})(e + 1) = \{\delta\}. \tag{10.4.5}$$

Edward chooses David, the least significant coworker, who was "excluded" from the graph of relations to make it decomposable. Note that removing d from the graph of relations does not change e's set of alternatives. The alternatives containing element δ can be chosen, and other subjects can incline subject e toward choosing them.

Let us now see what would happen if Alex and Bart did not try to persuade Edward of their disinterest and inclined him to choose them instead. Gregory, in this case, does not change his influence, and David's influence is not significant, because he is excluded from the graph of relations.

The new influences are as follows:

$$a = \{\alpha\}, \, b = \{\beta\}, \, g = \{\alpha, \beta, \gamma, \delta\} = 1. \qquad (10.4.6)$$

By substituting these values into (10.4.4), we find

$$e = \{\alpha, \beta\}. \qquad (10.4.7)$$

Edward chooses the alternative containing Alex and Bart, and then selects one of them. In this case, Alex and Bart have a chance of being nominated.

Chapter 11
Social processes and politics

In this chapter, we analyze such subjects as social groups, political parties, and criminal gangs. Practical use of the model in this area must be preceded by empirical study, which would identify units playing the role of subjects along with their sets of possible actions and the relations of conflict or cooperation between them.

11.1. Choice of economic system

Let us imagine a country where the choice of economic system is being discussed. There are forces preferring socialist economics and the nationalization of large private property. There are also opposing forces proposing economics based on a free market. Socialist economics will be designated α, and free market β. Thus, the universal set consists of two incompatible elements: α and β. The set of alternatives is

$$1 = \{\alpha, \beta\} - \text{not realizable},$$
$$\{\alpha\} - \text{choosing socialism (realizable)},$$
$$\{\beta\} - \text{choosing a free market (realizable)},$$
$$0 = \{\ \} - \text{inaction (realizable)}.$$

The social forces in the country are as follows:

political elite	a,
military	b,
secret police	c,
population	d,
business	e.

Suppose that the political elite is allied with the population and the secret police; the military forces are allied with population and in conflict with the political elite, the secret police, and business. The secret police are allied with the population, and business is in conflict with all others. These relations are depicted in Fig.11.1.1:

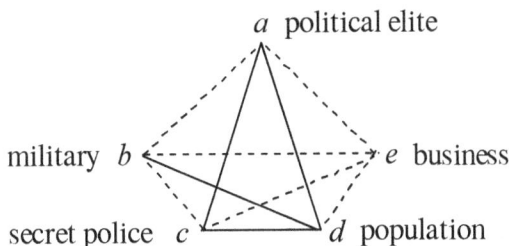

Fig. 11.1.1. Graph of relations

The political elite must make the decision concerning the choice of an economic system. Other social forces can influence the decision of the elite. The graph in Fig.11.1.1 is decomposable; it corresponds to the polynomial

$$e + d(b + ac) \qquad (11.1.1)$$

and the diagonal form

$$
\begin{array}{r}
[a]\,[c] \\
[b] + [ac] \\
[d]\,[b + ac] \\
[e] + [d(b + ac)] \\
[e + d(b + ac)]
\end{array}
\qquad (11.1.2)
$$

The equation for the political elite (a) is:

$$a = e + d\,(b + ac). \qquad (11.1.3)$$

Let the military demand that a certain course of economics development be chosen:

$$b = \overline{0} = \{\alpha, \beta\} = 1.$$

The secret police demands socialism:

$$c = \{\alpha\}.$$

The population is also willing to live under socialism:

$$d = \{\alpha\} \ .$$

Business demands a free market:

$$e = \{\beta\} \ .$$

We substitute these values into (11.1.3) and obtain

$$a = \{\beta\} + \{\alpha\}(1 + a\{\alpha\}) \qquad (11.1.4)$$

or

$$a = 1 = \{\alpha, \beta\}. \qquad (11.1.5)$$

Thus, under such a set of influences, the political elite chooses an active course; it may decide to build socialism or to develop a free-market economy; inaction is excluded.

Consider several other cases. For example, all social blocs are happy with the existing situation and do not want any change:

$$b = 0, \ c = 0, \ d = 0, \ e = 0.$$

We substitute these values into (11.1.3) and find that

$$a = 0.$$

If all blocs wish to have socialism,

$$a = \{\alpha\},$$

or capitalism

$$a = \{\beta\}.$$

In all these cases, the political elite obeys unanimous desire of the social blocs.

Let business (*e*) begin building affordable housing and clinics, create funds for the poor, etc., to attract the population (*d*). These measures change the population's desire to have socialism and incline the political elite to leave everything as is. The others' influence remains the same as in the first example. We substitute values $b = 1$, $c = \{\alpha\}$, $d = 0$, $e = \{\beta\}$ into equation (11.1.3) and obtain

$$a = \{\beta\} . \tag{11.1.6}$$

The change of these influences on the elite results in the choice of a free-market economy.

11.2. Appointment of a Prime Minister

The president (a) must appoint a Prime Minister. There are three candidates. α, β, γ, and six political parties, b, c, d, e, f, g. The universal set is $\{\alpha, \beta, \gamma\}$; the set of alternatives consists of eight elements.

Let the graph of relations be as follows:

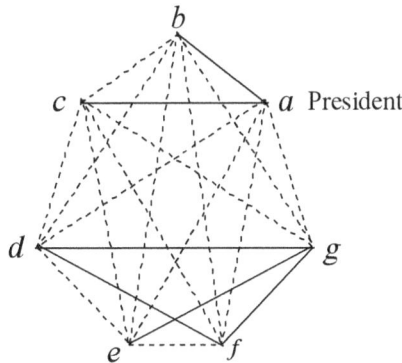

Fig. 11.2.1. Graph of relations

The subjects are organized in two groups: a, b, c and d, e, f, g, which are in conflict with each other. Inside the first group, a cooperates with b and c; b and c are in conflict. Inside the second group, g cooperates with d, e, and f; e is in conflict with d and f, who cooperate with each other. The graph in Fig. 11.2.1 corresponds to the polynomial

$$a(b + c) + g(e + df),\qquad\qquad (11.2.1)$$

and the diagonal form

$$
\begin{array}{l}
\qquad\qquad\qquad\qquad\qquad\qquad\qquad [d]\,[f]\\
\qquad\qquad [b] + [c]\qquad\qquad\qquad [e] + [df]\\
\qquad [a]\,[b + c]\qquad\qquad\quad [g]\,[e + df]\\
\quad [a(b + c)]\qquad\qquad\quad + [g(e + df]\\
[a(b + c) + g(e + df)]\qquad\qquad\qquad\qquad , \qquad\qquad (11.2.2)
\end{array}
$$

and the equation for a is

$$a = a(b + c) + g\,(e + df).\qquad\qquad (11.2.3)$$

Let party b incline the President to appoint α; party c inclines him not to appoint α, i.e., to appoint either β or γ. Parties d and f suggest appointing γ, party e wants the President to appoint β, and party g favors α. Thus, the values of the parties' influences on the President are as follows:

$$b = \{\alpha\},\; c = \{\beta,\,\gamma\},\; d = \{\gamma\},\; e = \{\beta\},\; f = \{\gamma\},\; g = \{\alpha\}.$$

Substitute these values into (11.2.3):

$$a = a\,(\{\alpha\} + \{\beta,\,\gamma\}) + \{\alpha\}\,(\{\beta\} + \{\gamma\}\{\gamma\}).\qquad\qquad (11.2.4)$$

After computation we obtain the equality

$$a = a.\qquad\qquad (11.2.5)$$

In this situation, the President has freedom of choice. He can appoint any candidate based on other factors; he even can refuse to make the appointment by choosing the empty alternative { }= 0.

What would happen if all parties insisted in appointing the same candidate, for example α? Assuming all variables, b, c, d, e, f, g, equal to $\{\alpha\}$ and substituting them into (11.2.3), we obtain

$$a = \{\alpha\},\qquad\qquad (11.2.6)$$

i.e., the President would appoint α. If all parties incline the

President not to appoint anyone, he will comply, because for the values of variables equal to { },

$$a = \{\ \}. \tag{11.2.7}$$

Consider some changes to the scenario. Parties d, e, f and g refuse to collaborate with the President. Party b continues to support α's appointment, and party c begins to support γ. Thus,

$$b = \{\alpha\},\ c = \{\gamma\}. \tag{11.2.8}$$

The new situation corresponds to graph

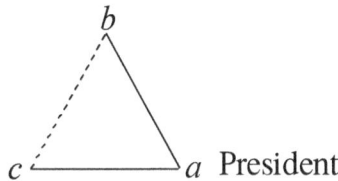

Fig. 11.2.2. Graph of relations

Find the polynomial

$$a(b + c), \tag{11.2.9}$$

and the diagonal form:

$$\begin{array}{c} [b] + [c] \\ [a]\,[b + c] \\ [a(b + c)] \end{array} \qquad , \tag{11.2.10}$$

and the equation for a:

$$a = a(b + c) + \bar{a}\ . \tag{11.2.11}$$

By substituting the values of influences from (11.2.8) into (11.2.11), we obtain

$$a = \{\alpha, \gamma\}a + \bar{a}\ . \tag{11.2.12}$$

In this equation, $A = \{\alpha,\ \gamma\}$, $B = 1$, i.e. $A \subset B$, thus, the equation does not have a solution, which means that the President is in a state of frustration and cannot make a choice.

11.3. Urban gangs

Five gangs operate in the city. Some of them cooperate, others are in conflict. Their activity has sharply increased. What is the reason? We assume that each gang faces a choice between active behavior (1) and passive behavior (0). We assume also that each gang inclines every other gang either toward activity or passivity. Suppose that after analyzing the situation we construct the following graph:

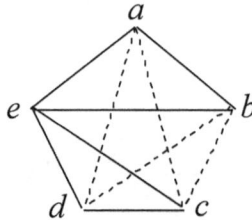

Fig. 11.3.1. Graph of gangs' relations

This graph corresponds to the polynomial

$$e(cd + ab) \tag{11.3.1}$$

and the diagonal form

$$
[e(cd + ab)]
\begin{array}{c}
{}_{[e]\,[cd\,+\,ab]} \\
{}_{[cd]}
\end{array}
\begin{array}{c}
{}^{[c]\,[d]} \\
\end{array}
+
\begin{array}{c}
{}^{[a]\,[b]} \\
{}_{[ab]}
\end{array}
\equiv 1. \tag{11.3.2}
$$

The group of gangs is in a superactive state, and each gang is also individually superactive. As was shown in Chapter 7, the state of superactivity does not depend on subjects' mutual influences; it depends on the persistence of the relations represented by the graph in Fig. 11.3.1. This is the reason the gangs' activity remains high.

The police trying to deal with the situation have only enough funds to neutralize the activity of one gang. For example, if the police neutralize gang a, the graph of relations becomes

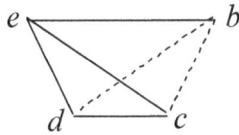

Fig. 11.3.2. Graph of relations after gang a is neutralized

This graph corresponds to the polynomial

$$e(b + dc) \qquad (11.3.3)$$

and the diagonal form

$$
\begin{array}{ll}
& [d]\,[c] \\
& [b] + [dc] \\
& [e]\,[b + dc] \\
[e(b + dc)] & \equiv 1. \qquad (11.3.4)
\end{array}
$$

A group consisting of four gangs, b, c, d, and e, remains superactive, so that the neutralization of gang a does not eliminate the gangs' superactivity.

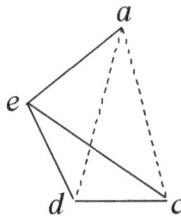

Fig. 11.3.3. Graph of relations after gang b is neutralized

How would the neutralization of gang b change the situation? The new graph of relations is given in Fig. 11.3.3. It corresponds to the polynomial

$$e(a + dc), \qquad (11.3.5)$$

and the diagonal form

$$
\begin{array}{c}
[d]\,[c] \\
[a] + [dc] \\
[e]\,[a + dc] \\
[e(a + dc)] \qquad\qquad\qquad \equiv 1.
\end{array}
\qquad (11.3.6)
$$

The group consisting of a, c, d, e is also superactive, so that neutralization of b will not diminish the activity of the others.

Similar analyses for gangs c or d demonstrate that neutralizing one of them leaves the group of remaining gangs superactive. What happens, however, if gang e, which cooperates with all other gangs, is neutralized? The new graph appears as follows:

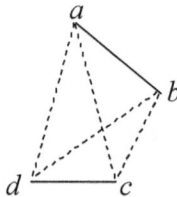

Fig. 11.3.4. Graph of relations after gang e is neutralized

It corresponds to the polynomial

$$
a\,b + c\,d \qquad (11.3.7)
$$

and the diagonal form

$$
\begin{array}{c}
[a]\,[b] \qquad [d]\,[c] \\
[ab] \qquad\quad + [dc] \\
[ab + dc] \qquad\qquad\qquad \not\equiv 1
\end{array}
\qquad (11.3.8)
$$

that is, not identically equal to 1. For example, for $a = 0$, $b = 0$, $c = 0$, $d = 0$, its value is 0. Thus, the neutralization of gang e makes the state of the group of gangs not superactive. So, within the model's framework, the neutralization of gang e will diminish the level of criminal activity in the city.

Chapter 12
International relations

In modeling international relations, the subjects under consideration are countries or groups of countries. We assume that in a situation of international crisis, each subject faces a choice of types of behavior: active (1) or passive (0). In this chapter, we will analyze several actual international crises. We will use concepts of cooperation, conflict, and influence. We do not have objective measuring devices to determine relations and influences without ambiguity. The only available measurement is expert evaluation. In all the examples analyzed in this book, the expert is the author. Having certain knowledge about real subjects and their interactions, I use this knowledge in my evaluations. For example, it is obvious that during the Second World War, Germany and the Soviet Union were in conflict. It is less obvious, but also reasonable to believe that during the prewar months these countries were not in conflict. To evaluate their relations on the bipolar construct cooperation-conflict, one would use the term "cooperation." Other analysts may disagree. Ideally, we would stand at a white board and model the prewar situation together. As my experience tells me, the process of modeling by means of varied initial data is an effective analytical instrument. The model puts the existing information in logical order and stimulates intuitive insight.

12.1. Europe, 1941

Let us analyze international relations during the first half of 1941. The critical relations in this period were between Germany, Soviet Union, England, and USA. Prior to Germany's attack on Soviet Union, the situation can be represented by the graph in Fig. 12.1.1:

USA a ┌─────────┐ b Soviet Union

England c └─────────┘ d Germany

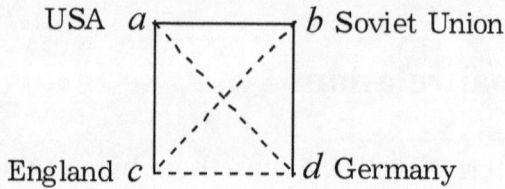

Fig. 12.1.1. Relations in 1941 prior to June 22
(first version)

Although relations between the USA and the Soviet Union were not very friendly, both parties viewed each other as a potential ally in the fight against Hitler. Then Germany and Soviet Union signed a pact of non-aggression. Hitler kept assuring Stalin that the concentration of German troops on the Soviet borders was just a distracting maneuver in the fight against England, and Stalin continued to supply raw materials to Hitler. So, in the framework of our model, we consider that Germany and Soviet Union were not in a relation of conflict. Since evaluations according to our model are binary, we say that they cooperated. England was in cooperation with USA; this alliance was vitally important for England in its struggle against Germany. Relations between England and the Soviet Union were not good: Stalin believed that England had pressured Hitler to attack the Soviet Union, and the English government saw that Stalin was helping Hitler economically. The USA supported England and was in conflict with Germany.

The graph in Fig. 12.1.1 is not decomposable, because it is $S_{(4)}$. Let us analyze each country's choices.

USA. The order of significance is as follows: England, Germany, Soviet Union. After removal of node b, the graph becomes

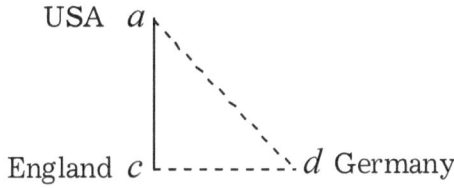

USA a

England c — — — — — — d Germany

Fig. 12.1.2. Graph of relations after Soviet Union is removed

This graph corresponds to the polynomial

$$d + ac, \qquad (12.1.1)$$

and the diagonal form

$$\begin{array}{c} [a]\,[c] \\ [d] + [ac] \\ [d + ac] \end{array} \qquad , \qquad (12.1.2)$$

and the equation for a is

$$a = d + ac. \qquad (12.1.3)$$

Germany wants the USA to remain passive

$$d = 0,$$

and England pressures the USA to defend it actively against Germany

$$c = 1.$$

Equation (12.1.3) becomes

$$a = a. \qquad (12.1.4)$$

Therefore, during this period, United States has freedom of choice.

Soviet Union. The order of significance is as follows: Germany, USA, England. The graph of relations after England's removal is given in Fig.12.1.3:

USA *a* ──────────┐ *b* Soviet Union

d Germany

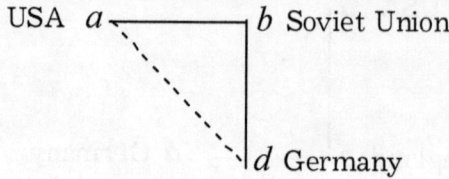

Fig. 12.1.3. Graph of relations after removal of England

This graph corresponds to the polynomial

$$b(a + d), \qquad (12.1.5)$$

and the diagonal form

$$[a] + [d]$$
$$[b] [a + d]$$
$$[b(a + d)] \qquad (12.1.6)$$

and the equation for *b* is:

$$b = (a + d)b + \bar{b} . \qquad (12.1.7)$$

Both the USA and Germany incline Stalin toward a passive line of behavior

$$a = 0, d = 0.$$

Thus equation (12.1.7) appears as

$$b = \bar{b} . \qquad (12.1.8)$$

Therefore, the Soviet leadership is in a state of frustration; this corresponds to Stalin's hesitations and lack of initiative in the months preceding the German invasion. He did not appear as an aggressor in the eyes of the USA and England and was afraid to be blamed for provoking Germany into a conflict.

Germany. The order of other countries' significance is: Soviet Union, England, USA. After the removal of the USA the graph becomes

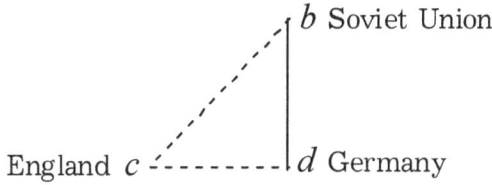

Fig. 12.1.4. Graph of relations after removal of USA

This graph corresponds to the polynomial

$$c + bd, \tag{12.1.9}$$

and the diagonal form

$$\begin{array}{c} [b]\,[d] \\ [c] + [bd] \\ [c + bd] \end{array} \qquad , \tag{12.1.10}$$

and the equation for d is:

$$d = c + bd. \tag{12.1.11}$$

England appears inaccessible to Hitler and as such stimulated him to action against the Soviet Union, since the latter appears weak due to the Kremlin's indecisive policy. Thus, $c = 1$ and equation (12.1.11) becomes

$$d = 1. \tag{12.1.12}$$

Germany, during this period, followed an active line of behavior.

England. The order of significance is: Germany, USA, Soviet Union. After the removal of the Soviet Union, the graph of relations became as given in Fig.12.1.2. The equation for England is

$$c = d + ac. \tag{12.1.13}$$

Germany carried out military actions against England stimulating it to activity: $d = 1$. Equation (12.1.13) appears now

$$c = 1. \tag{12.1.14}$$

England is in an active state.

With all the given assumptions, in the period from January 1941 to June 22 the countries' states are as follows: Germany and England are active, Soviet Union is in frustration, USA has freedom of choice.

As was mentioned in the introduction to this chapter, ascribing 'cooperation' to German-Soviet relations during the prewar months is not completely obvious. Let us now ask what predictions would be given by the model, if we supposed that Germany and the Soviet Union were in conflict in that period. The graph of such relations is depicted in Fig.12.1.5.

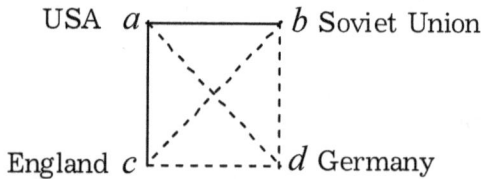

USA a ——— b Soviet Union

England c - - - - - - d Germany

Fig. 12.1.5. Relations in 1941 prior to June 22
(second version)

This graph corresponds to the polynomial

$$d + a(b + c) \tag{12.1.15}$$

and the diagonal form

$$[b] + [c]$$
$$[a]\,[b + c]$$
$$[d] + [a(b + c)]$$
$$[d + a(b + c)] \tag{12.1.16}$$

After transformation we obtain the equation for Germany:

$$d = d + a\overline{d} \ . \tag{12.1.17}$$

Solutions to this equation are given by the inequalities

$$1 \supseteq d \supseteq a \ . \tag{12.1.18}$$

We see that Germany's choice depends neither on Soviet influence nor on England's influence. It depends only on American influence. If the USA inclines Germany toward activity ($a = 1$), Germany will choose an active line of behavior, and if USA inclines Germany to passivity, it will have freedom of choice and may choose a passive line of behavior. Such significant German dependence on US influence, combined with the absence of any dependence on England or the Soviet Union, seems unlikely, from a historical point of view. So, the second version gives a less realistic prediction. It is based on the less likely assumption that the Soviet Union and Germany were in conflict during that period.

12.2. Hungary, 1956

The Budapest student demonstration that began on October 23, 1956, developed quickly into a general uprising against the political system established in Hungary by the Soviet Union. On October 24, Imre Nad became the leader of the revolt and the head of the country. On November 4, Soviet troops entered Budapest and suppressed the rebellion. Imre Nad was arrested and executed two years later for "high treason."

Western reaction to the "Hungarian events" was quite passive. NATO troops were not sent to Hungary; the USA produced nothing more than an accusatory declaration. The Soviet Union felt itself to be the master of the situation.

In modeling international relations during the Hungarian uprising, we will consider four subjects: Hungary, Soviet Union, USA, and Western Europe. The three subjects: Hungary, Western Europe, and the USA were in cooperation, at least morally. Soviet Union was in conflict with all of them. The graph of their relations is given in Fig. 12.2.1:

USA a - - - - - - - c Soviet Union

Hungary b d Western Europe

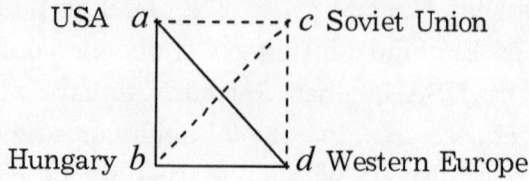

Fig. 12.2.1. Graph of relations during the Hungarian revolt

This graph corresponds to the polynomial

$$c + abd, \tag{12.2.1}$$

and the diagonal form

$$[a]\,[b]\,[d]$$
$$[c] + [abd]$$
$$[c + abd] \tag{12.2.2}$$

and the equation for the subjects is:

$$x = c + abd, \tag{12.2.3}$$

where $x = a, b, c, d$.

Let us construct the matrix of influences:

Table 12.2.1.
Matrix of influences

		a	b	c	d
USA	a	a	1	0	0
Hungary	b	1	b	1	1
Soviet Union	c	0	1	c	0
Western Europe	d	0	0	0	d

The USA appeals to the Soviet Union to be passive; it does not stimulate Western Europe to action, but supports the revolt in public declarations. Soviet Union tries to calm the USA and Western Europe, but at the same time stimulates Hungary to activity by its aggression. Hungary appeals to the USA and Western Europe to become active and, by the fact of its uprising, stimulates the Soviet Union to activity. Western Europe stimulates all parties to choose a passive line of behavior.

Substitute the values of a, b, c, d from Table 12.2.1 into equation (12.2.3):

$$a = 0 + a\ 1\ 0 = 0,$$
$$b = 1 + 1\ b\ 0 = 1,$$
$$c = c + 0\ 1\ 0 = c,$$
$$d = 0 + 0\ 1\ d = 0.$$

The model shows that the USA and Western Europe choose a passive line of behavior, Hungary chooses activity, and Soviet Union has freedom of choice.

12.3. Iranian crisis, 2006

This crisis, which reached its peak in June, 2006, was connected with international apprehension over Iran's production of enriched uranium, usable for nuclear weapons. The subjects involved in this crisis were divided in two groups: Iran, China and Russia (the last two supporting Iran) on the one side, and USA, Israel and European Union (adversarial to Iran) on the other. The subjects constituting each group cooperated with each other and were in conflict with the subjects in the other group, except for European Union and Russia, who were in friendly relations at that time. The situation described corresponds to the following graph (Fig.12.3.1):

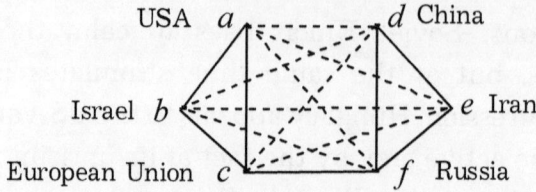

Fig. 12.3.1. Relations between the main subjects involved
in the Iranian crisis of 2006

This graph is not decomposable, because its subgraph $<a, c, d, f>$
is $S_{(4)}$.

USA. We assume that the order of other subjects'
significance for USA is as follows: Iran, Israel, European Union,
Russia, China. Node d is removed first, which results in the graph
in Fig. 12.3.2:

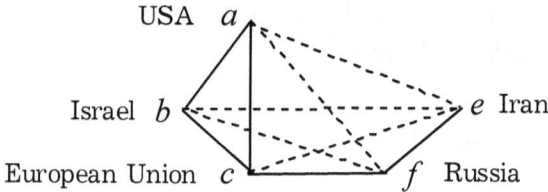

Fig. 12.3.2. Graph of relations after China is removed

This graph is not decomposable either, because its subgraph $<a, c,$
$f, e>$ is $S_{(4)}$. Next, node f corresponding to Russia is removed, and
a new graph appears (Fig. 12.3.3):

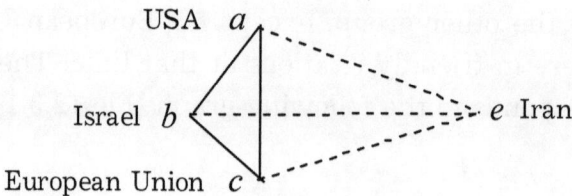

Fig. 12.3.3. Graph of relations after China and Russia are removed

This graph is decomposable. It corresponds to the polynomial

$$e + abc, \qquad (12.3.1)$$

and the diagonal form

$$[a]\,[b]\,[c]$$
$$[e] + [abc]$$
$$[e + abc] \qquad\qquad , \qquad (12.3.2)$$

and the equation for the USA is

$$a = e + abc . \qquad (12.3.3)$$

In the framework of this situation, Iran inclines USA to activity by the fact that it is preparing to become a nuclear power, so $e = 1$. This information is enough to write

$$a = 1 + abc = 1, \qquad (12.3.4)$$

i.e., the United States chooses an active line of behavior.

Israel. The order of other subjects' significance is as follows: Iran, United States, Russia, European Union, China. First, node d (China) is removed and we obtain the graph in Fig.12.3.2, which is not decomposable. Next, node c corresponding to the European Union is removed:

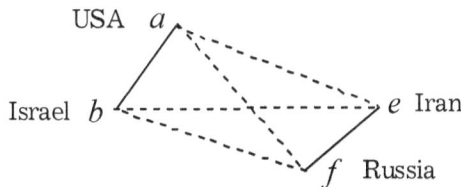

Fig. 12.3.4. Graph of relations after China and European Union are removed

This graph corresponds to polynomial (12.3.5):

$$ab + ef, \qquad (12.3.5)$$

and to the diagonal form

$$
\begin{array}{cc}
[a]\,[b] & [e]\,[f] \\
[ab] & +\,[ef] \\
[ab + ef] &
\end{array}
\qquad , \qquad (12.3.6)
$$

and the equation for Israel is

$$b = ab + ef.$$

The United States inclines Israel to passive behavior, fearing that the situation may become explosive ($a = 0$). Russia also inclines Israel to passivity, fearing a possible nuclear strike against Iran by US and Israeli forces ($f = 0$). This information is enough to determine Israel's choice:

$$b = 0, \qquad (12.3.7)$$

i.e., Israel chooses a passive line of behavior.

    European Union. The order of significance is: USA, Iran, Russia, Israel, China. After node d (China) is removed, the graph (Fig.12.3.2) remains non-decomposable. Then node b (Israel) is removed, resulting in graph (Fig. 12.3.5):

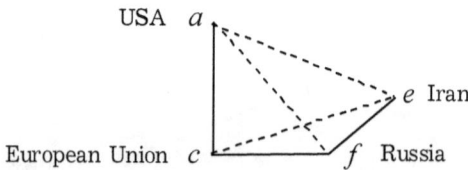

Fig. 12.3.5. Graph of relations after removal China and Israel

This graph is not decomposable, because it is $S_{(4)}$. In the next step, node f (Russia) is removed. The new graph (Fig. 12.3.6) is decomposable. It corresponds to the polynomial

$$e + ac, \qquad (12.3.8)$$

and the diagonal form

$$[a]\,[c]$$
$$[e] + [ac]$$
$$[e + ac] \qquad\qquad . \qquad\qquad (12.3.9)$$

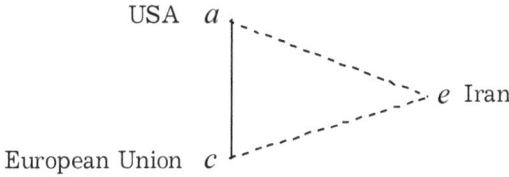

Fig. 12.3.6. Graph of relations after China, Israel, and Russia are removed

The equation for the European Union is

$$c = e + ac. \qquad\qquad (12.3.10)$$

Iran inclines the European Union toward passivity ($e = 0$), but the US urges the European Union to activity ($a = 1$). As a result, we obtain the equation

$$c = c. \qquad\qquad (12.3.11)$$

Thus, the European Union has freedom of choice.

Russia. The order of significance is as follows: USA, Iran, China, European Union, Israel. Removing Israel (node b) results in the graph given in Fig.12.3.7. The next to be removed is the European Union (node c). This results in graph (Fig. 12.3.8) which is decomposable. It corresponds to the polynomial

$$a + def \qquad\qquad (12.3.12)$$

and the diagonal form

$$[d]\,[e]\,[f]$$
$$[a] + [def]$$
$$[a + def] \qquad\qquad , \qquad\qquad (12.3.13)$$

and the equation for Russia is

$$f = a + def. \tag{12.3.14}$$

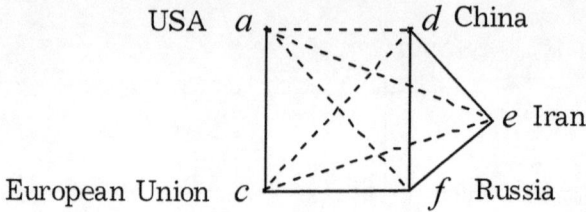

Fig. 12.3.7. Graph of relations after Israel is excluded

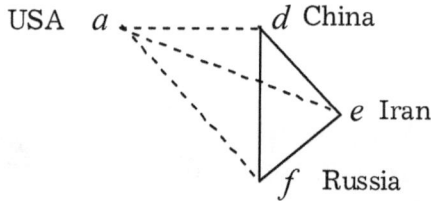

Fig. 12.3.8. Graph of relations after Israel and European Union
are excluded

The US inclines Russia toward passivity: $a = 0$; Iran and
China push Russia toward action in defense of Iran: $e = 1$, $d = 1$.
Thus,

$$f = f , \tag{12.3.15}$$

Russia has freedom of choice.

China. The order of significance is: USA, Russia, Iran,
European Union, Israel. China's list of significance ends with the
European Union and Israel, as does Russia's list. This means that
China operates with the graph in Fig. 12.3.8. The equation for
China is

$$d = a + def. \tag{12.3.16}$$

The US inclines China toward passivity ($a = 0$), Iran toward activity ($e = 1$), Russia does not take steps to induce China to actively support Iran's right to conduct uranium enrichment, so we consider that Russia inclines China toward passivity: $f = 0$. Therefore,

$$d = 0, \qquad (12.3.17)$$

China chooses a passive line of behavior.

Iran. The order of other countries' significance for Iran is as follows: USA, Israel, European Union, Russia, China. After excluding China and Russia, the graph of relations becomes as given in Fig. 12.3.3. The equation for Iran is:

$$e = e + abc. \qquad (12.3.18)$$

European Union inclines Iran toward passivity: $c = 0$. Independently from the values of variables a and b the equation $abc = 0$ holds, and the knowledge of the European Union's influence is enough to find

$$e = e . \qquad (12.3.19)$$

Thus, Iran has freedom of choice.

Here are the results of our analysis. During the Iranian crisis of 2006, three subjects had clear political lines: USA active, Israel and China passive. Three other subjects (Iran, Russia, and European Union) had less clearly defined policies. The model ascribed freedom of choice to these subjects.

12.4. The analysis of frustration

In this section, we will demonstrate which political processes in international relations may correspond to a state of frustration.

For a period of several years, the Palestinian terrorist organization Hamas has shelled Israeli territory. During the last few days of December, 2008, Israel attacked terrorist bases in Gaza. We will consider the situation prior to the entry of land forces into Gaza. It is depicted by the graph in Fig. 12.4.1. Note that the subjects in this situation are of different types. The US and Israel are countries, Hamas is a terrorist organization, and the anti-Israeli forces are a group of countries and organizations acting against Israel. The USA cooperates with Israel and, in general, with anti-Israel forces, but is in conflict with Hamas. Israel is in conflict with Hamas and with the anti-Israel forces; the latter cooperate with Hamas. We will model the US choice.

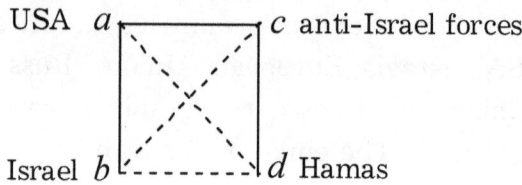

USA a c anti-Israel forces

Israel b d Hamas

Fig. 12.4.1. Graph of relations during Israel - Hamas conflict

This graph is not decomposable, because it is $S_{(4)}$. We assume that for the US the order of significance of the subjects involved in this crisis is as follows: Israel, anti-Israel forces, Hamas. So, Hamas is excluded. The remaining three parties correspond to graph $<a, b, c>$, with polynomial:

$$a(b + c)$$

and diagonal form:

$$[b] + [c]$$
$$[a]\,[b + c]$$
$$[a(b + c)] \qquad\qquad . \qquad\qquad (12.4.1)$$

The equation for the US is

$$a = (b + c)a + \bar{a} \ . \qquad\qquad (12.4.2)$$

Both Israel and anti-Israeli forces do not want US to be active, but their reasons are different. Israel does not want the US to stop the destruction of terrorist bases and return to fruitless process of negotiations. The anti-Israeli forces are afraid that the US may assist Israel militarily and politically. So, both subjects incline the USA to passivity: $b = 0$, $c = 0$. Equation (12.4.2) becomes

$$a = \bar{a}. \tag{12.4.3}$$

The model predicts that the US is in a state of frustration. How is this manifested? On December 31, 2008, *The Los Angeles Times* published the article "Behind closed doors, US seeks an exit," by Paul Richter. He wrote about the conflict between the positions of the White House and the State Department concerning the Israel - Hamas conflict. While President Bush supported Israel's right to defend itself, the State Department considered international pressure against the military campaign. We see that the influence of Israeli and anti-Israel forces generated two opposite tendencies, which resulted in the different positions of the White House and the State Department. For this reason, the US could not choose a line of behavior during the first days of the conflict.

Chapter 13
Military decisions

In this chapter, we will show how the theory of reflexive games can be used for predicting decisions made on the battlefield. First, we will examine the theory by comparing its results with obvious intuitive predictions; then, we will model situations of military interaction.

13.1. Intuition and predictions by the model

A battalion commander has received an order to assemble troops near a river and, if possible, to force the river crossing. Three crossings exist in the area of the battalion's location: A, B, C. Each of them is defended by an enemy platoon, which constitutes a serious obstacle. What decision will the commander make? Will he cross the river at point A, point B, or point C, or will he not cross at all? To answer this question, we have only the brief description given above. So, we must look for other constraints on the commander's choice. We cannot find any, because there is not enough information. This means that the subject's choice depends on factors not included in the description, and we cannot predict his choice; i.e., from our position, the battalion commander has *freedom of choice*.

Suppose that we construct a model based on the given situation, and it predicts that the commander will choose to cross the river at point A. This prediction would undermine our trust in the model, because in the description there are no differences between A, B and C. We will consider a prediction satisfactory only if it states that the commander can choose any crossing, i.e., that he has freedom of choice.

Check now what our model predicts. The following graph corresponds to the situation (Fig.13.1.1):

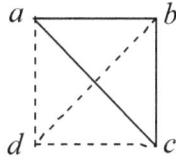

Fig. 13.1.1. Graph of relations

Node *d* depicts the battalion, nodes *a*, *b* and *c* depict platoons defending crossings A, B and C, respectively. The graph in Fig. 13.1.1 corresponds to the polynomial

$$d + abc \qquad (13.1.1)$$

and to the diagonal form

$$[a]\,[b]\,[c]$$
$$[d] + [abc]$$
$$[d + abc] \qquad (13.1.2)$$

The equation for the battalion commander is

$$d = d + abc. \qquad (13.1.3)$$

The universal set consists of three actions: α, β, γ, where α is to cross the river at point A, β to cross at point B, and γ to cross at point C. Only $\{\alpha\}$, $\{\beta\}$, $\{\gamma\}$ and 0 are realizable alternatives. The choice of 0 means that the commander makes the decision not to force the river crossing. Each enemy's platoon influences the commander not to use the crossing that the platoon defends:

$$a = \overline{\{\alpha\}} : \text{do not use crossing A,}$$
$$b = \overline{\{\beta\}} : \text{do not use crossing B,}$$
$$c = \overline{\{\gamma\}} : \text{do not use crossing C,}$$

or

$$a = \{\beta, \gamma\},$$
$$b = \{\alpha, \gamma\},$$
$$c = \{\alpha, \beta\}.$$

We substitute these values into (13.1.3) and obtain

$$d = d + \{\beta, \gamma\}\{\alpha, \gamma\}\{\alpha, \beta\}. \qquad (13.1.4)$$

Since

$$\{\beta, \gamma\}\{\alpha, \gamma\}\{\alpha, \beta\} = 0,$$

$$d = d, \qquad (13.1.5)$$

i.e., the battalion commander has freedom of choice. We see that our model's prediction coincides with the conclusion made after intuitive analysis of the situation.

Suppose that the enemy fortifies the defense of crossings B and C, but leaves A's defense as is. What decision will be made by the commander under this condition? It is clear that crossing at A becomes more attractive than B and C, but the potential use of B and C must remain.

Let us model this situation. First, we determine the influences. Platoon a, which defends crossing A, does not repel the battalion, but attracts it:

$$a = \{\alpha\}. \qquad (13.1.6)$$

The influences of platoons b and c remain as they were previously:

$$b = \{\alpha, \gamma\}, \qquad (13.1.7)$$

$$c = \{\alpha, \beta\}. \qquad (13.1.8)$$

We substitute these values into (13.1.3) and obtain:

$$d = d + \{\alpha\}\{\alpha, \gamma\}\{\alpha, \beta\}. \qquad (13.1.9)$$

Since

$$\{\alpha\}\{\alpha, \gamma\}\{\alpha, \beta\} = \{\alpha\}, \qquad (13.1.10)$$

equation (13.1.9) becomes

$$d = d + \{\alpha\}, \qquad (13.1.11)$$

or

$$d = d + \{\alpha\}\bar{d} . \qquad (13.1.12)$$

The solutions of this equation are given by the inequalities

$$1 \supseteq d \supseteq \{\alpha\}. \qquad (13.1.13)$$

Thus,

$$d = \{\alpha, \beta, \gamma\}, \{\alpha, \beta\}, \{\alpha, \gamma\}, \{\alpha\}. \qquad (13.1.14)$$

The battalion commander can choose one of four alternatives, each of which contains action α. There is no other action that is included in all alternatives. So, crossing at A is *singled out*. With the choice of any alternative, the battalion commander can use this particular crossing.

We see that, in simple cases, our model's predictions correspond to those of intuition and common sense.

13.2. Choosing a path

Consider a more complicated case. The separate battalion d has the goal of descending from the mountains into the valley. The enemy's platoon, a, tries to hold the battalion in the mountains. All routes into the valley go through villages b and c. The villages' inhabitants are hostile to the battalion and support the enemy. Moreover, they are in conflict with each other. The situation corresponds to the graph in Fig.13.2.1.

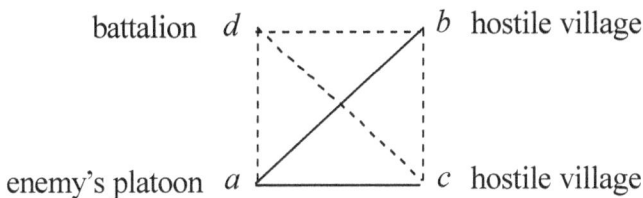

battalion d b hostile village

enemy's platoon a c hostile village

Fig. 13.2.1. Graph of relations

The universal set of actions for the commander contains all routes descending from the mountains into the valley. We assume that they are incompatible, i.e., the battalion can move along one route only. The set of alternatives M consists of all subsets of the universal set including the empty set, which corresponds to a refusal to make a descent. Set a consists of routes which the enemy's behavior inclines the battalion commander to use. For example, some routes are not covered by the enemy, which may incline the commander to use them. Sets b and c contain routes, which the villages' inhabitants incline the commander to use. The graph in Fig.13.2.1 is decomposable; it corresponds to the polynomial

$$d + a(b + c) \qquad (13.2.1)$$

and the diagonal form

$$\begin{array}{l} [b] + [c] \\ [a]\,[b + c] \\ [d] + [a(b + c)] \\ [d + a(b + c)] \end{array} \qquad (13.2.2)$$

The equation for the battalion commander is

$$d = d + a . \qquad (13.2.3)$$

There are no variables b and c in this equation, so that, the commander's choice does not depend on the hostile villages' inhabitants. By solving (13.2.3), we find

$$1 \supseteq d \supseteq a . \qquad (13.2.4)$$

When a is not empty, the battalion commander can choose any alternative containing a as a subset, and after that, he can realize one of the subsets of this alternative, which is a particular route. When $a = 0$, i.e., the enemy does not pressure the commander toward any route, the commander has freedom of choice and can choose any alternative, including the empty one. If the chosen alternative is not empty, the commander can single out a route for the battalion to descend to the valley.

What will be the commander's choice if the inhabitants of villages b and c are in cooperation instead of conflict? The graph of the relations in this case is given in Fig. 13.2.2.

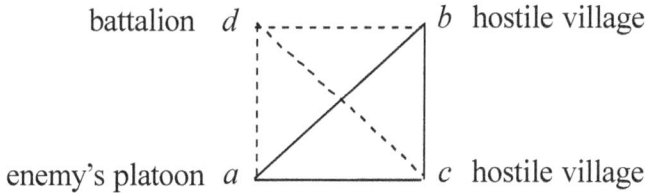

battalion d ············ b hostile village

enemy's platoon a c hostile village

Fig. 13.2.2. Graph of relations

This graph corresponds to the polynomial

$$d + abc \tag{13.2.5}$$

and the diagonal form

$$[a\} [b] [c]$$
$$[d] + [abc]$$
$$[d + abc] \tag{13.2.6}$$

The equation for the battalion commander is

$$d = d + abc, \tag{13.2.7}$$

hence,

$$1 \supseteq d \supseteq abc . \tag{13.2.8}$$

Thus, when the inhabitants of the hostile villages a and b cooperate with each other, the commander's choice depends on their influence.

For example, village b inclines the battalion commander to choose alternative 0 (b=0). Pressure toward zero means that subject b, by his behavior, prompts the battalion to inaction. As a result, the battalion commander obtains freedom of choice.

13.3. Reflexive control

Consider the scenario corresponding to Fig.13.2.1. The commander's choice is given by the inequalities $1 \supseteq d \supseteq a$. Let the head of the enemy platoon decide to ambush the battalion on one of the routes and to use reflexive control as a tool. If he realizes influence $a = 1$ and persuades the battalion commander that every route to the valley is safe, he will not be able to predict the battalion's route and will not be able to set the ambush. If the head of the enemy unit realizes influence $a = 0$, i.e., all routes down are dangerous, then the battalion commander will obtain freedom of choice, and it will again be impossible to predict his route. The best expedient is to pressure the commander toward choosing a concrete route λ, i.e., to use influence $a = \{\lambda\}$. Then the battalion commander's choice will be given by the inequalities

$$1 \supseteq d \supseteq \{\lambda\}, \qquad\qquad (13.3.1)$$

which means that the choice will be made from a set of subsets, each of which contains λ. With this reflexive control, route λ will be included in each alternative. There is no guaranty, however, that the battalion will use this route.

Chapter 14
Theorem on jurisprudence

A verdict of an ideal judge must not depend on the emotional influences of other members of judicial process. He must be neither captivated by the defender's eloquence nor terrified by the prosecutor's accusations. An ideal judge's decision is based only on established facts and the law. In this chapter we will show how the model of an ideal trial can be constructed with the help of the reflexive game theory.

14.1. Ideal trial

According to the traditional arrangement, four subjects participate in a trial: the defendant, the judge, the defender, and the prosecutor. The universal set of choice for the judge consists of two actions: to make the decisions 'guilty' (α) or 'not guilty' (β). So, there are four alternatives, represented as a Boolean lattice in Fig. 14.1.1:

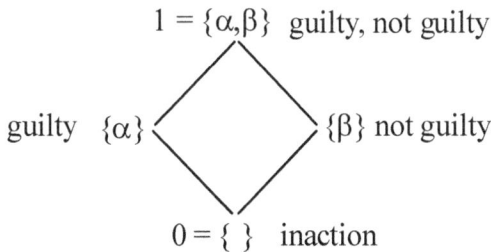

$$1 = \{\alpha, \beta\} \quad \text{guilty, not guilty}$$

guilty $\{\alpha\}$ $\{\beta\}$ not guilty

$$0 = \{\ \} \quad \text{inaction}$$

Fig. 14.1.1. A judge's alternatives for choice

The members of the trial may incline the judge to choose one or another alternative by exercising emotional pressure. If a judge chooses $\{\ \}$, it means that he refuses to function. If a judge chooses $\{\alpha, \beta\}$, it means that he can realize either $\{\alpha\}$, or $\{\beta\}$, i.e., declare the defendant either guilty or not guilty based on rational arguments. We will call the trial *perfect*, if the judge chooses alternative $1 = \{\alpha, \beta\}$ under any emotional pressure from the other members of the trial: prosecutor, defendant, and defender.

The theorem on jurisprudence. The classic trial is perfect.

Proof. The graph in Fig. 14.1.2 depicts relations between the members of a classic trial. The defendant and defender cooperate with each other and are in conflict with the prosecutor. The judge is in a relation of collaboration (cooperation, in our terminology) with all others.

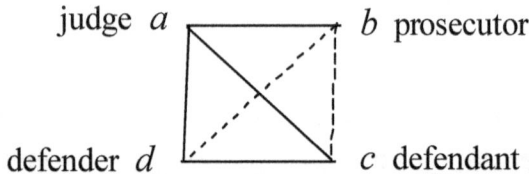

judge a b prosecutor

defender d c defendant

Fig. 14.1.2. Graph of relations in a classic trial

This graph corresponds to the polynomial

$$a(b + cd) \tag{14.1.1}$$

and the diagonal form

$$[c]\,[d]$$
$$[b] + [cd]$$
$$[a]\,[b + cd]$$
$$[a(b + cd)] \qquad\qquad \equiv 1 = \{\alpha, \beta\}. \quad \square \tag{14.1.2}$$

We see that the judge is superactive; this means that he makes his decisions independently of any influence from other members of the trial. For example, suppose all of them - defendant, prosecutor, and defender – are eagerly persuading the judge that the defendant is guilty: $c = \{\alpha\}$, $b = \{\alpha\}$, $d = \{\alpha\}$. In this case, according to equation (14.1.2), the judge will still make an independent decision.

14.2. Trial without defender

Will a trial remain perfect, if the defender is removed from the scheme? In this case, the graph of relations becomes as follows:

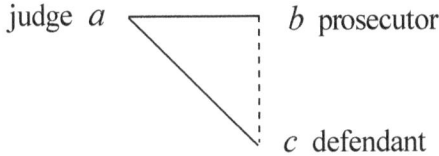

judge a ———————— b prosecutor

c defendant

Fig. 14.2.1. Graph of relations after removal of defender

This graph corresponds to the polynomial

$$a(b + c) \tag{14.2.1}$$

and the diagonal form

$$\begin{array}{c} [b] + [c] \\ [a]\,[b + c] \\ [a(b + c)] \end{array} \tag{14.2.2}$$

The equation for the judge is

$$a = (b + c)a + \bar{a} \,. \tag{14.2.3}$$

This equation has no solution if

$$(b + c) \subset 1 \,. \tag{14.2.4}$$

For example, if $b = \{a\}$ and $c = \{a\}$, i.e., the prosecutor states that the defendant is guilty, and the defendant makes a 'guilty' plea, equation (14.2.4) holds, the judge cannot make a decision, he is in a state of frustration. Therefore, a trial without the defender is not perfect.

The relations graph in Fig. 14.1.2 is the result of centuries-old traditions. We can state with certainty that the ideal judge in the ideal trial will give a verdict independently of any emotional pressure from other members of the judicial procedure.

Conclusion

Classical game theory involves problems of two types: descriptive and prescriptive. The former is related to choice prediction, and the latter indicate the choices the player must make. The principle of max-min is used in both cases to minimize losses. By using game theory we agree implicitly that the max-min principle describes the process of players' decision making and we also base our own decisions on this principle. If we believe that other players are inclined to irrational risk and we also are willing to risk, we must not use classical game theory to make decisions.

The theory of reflexive game theory is based on the model of mental mechanisms realizing the anti-selfishness principle. This theory is also capable of solving descriptive and prescriptive problems. In the former case, we can predict subjects' choices, assuming that they obey the anti-selfishness principle. In the latter case, knowing the relations between the subjects and the influences on them, we can find if there is a choice satisfying the anti-selfishness principle and what it is.

The absence of a choice signals to us that we have to change our view of the problem. Consider an example. I have two friends living in a town far from me. They have in bad terms. I may go to their town and stay with one of them. Each friend wants me to visit but does not invite me, because he is afraid that I would stay with the other. No hotel or motel exists in that town. A formal analysis of this situation shows that I am in a state of frustration, i.e., I cannot decide either to go or not to go.
The impossibility of making choice that satisfies the anti-selfishness principle generates the following question: is it worth making this particular choice? Perhaps, I'd better consider another alternative, like inviting my friends to visit me.

Therefore, the theory of reflexive games is capable of guiding our choices and of indicating situations a fresh recasting of the alternatives is needed.

Appendix

Problems and Exercises

Introduction

The brief descriptions of subjects' behavior are given below. In each case determine whether the subjects' actions satisfy the anti-selfishness principle.

1. In 356 b. a. Herostratus sets fire to the temple of Artemis in his quest for fame.

2. A rich man leaves his fortune to the church in hope of getting to heaven.

3. Children ask their father to come home early from work. The father comes late because he stopped at a bar.

4. The people of a kingdom ask the king to declare amnesty, but the king remains unbending in spite of the risk of uprising.

5. Danko, a legendary hero, tears out his heart to light the way through forest for his people.

6. People ask for bread and circuses from the ruler, but he gives them only circuses, because he too loves them much.

7. In spite of the spectators' protests, the emperor pardons a defeated gladiator because he does not want to kill people.

8. The next time he pardons a defeated gladiator, the emperor tells the spectators that anyone who desires death for another will go to hell. The silence shows that the people understand.

Chapter 1

I. Draw Venn diagrams for the following sets, assuming that they have non-empty intersections.

1. $A + BC$ 2. $(A + B)C$ 3. ABC 4. $A + B + C$

5. $\overline{A + B}$ 6. $\overline{AB}$ 7. $A + \overline{B}$ 8. $A\overline{B}$

9. $\overline{A} + \overline{B}$ 10. $\overline{\overline{A} + \overline{B}}$ 11. $\overline{A}\ \overline{B}$ 12. $\overline{\overline{A} + B}$

II. Analyze each of the following equations:
     (1) determine whether it can be solved
     (2) if it is, indicate an interval

13. $x = (a + b)x + b\overline{x}$

14. $x = bx + (a + b)\overline{x}$, where $(a + b) \supset b$

15. $x = x + a\overline{x} + b$ 16. $x = x + b$ 17. $x = \overline{x} + c$

III. Analyze the equations defined on the set of all subsets of the universal set $\{\alpha, \beta, \gamma\}$:

18. $x = \{\alpha, \beta\}\, x + \{\alpha\}$ 19. $x = \{\beta\}\, x + \{\alpha, \beta\}\overline{x}$

20. $x = \{\alpha\}\, x + \{\alpha\}\overline{x}$ 21. $x = \overline{\{\alpha\}}\, x + \{\alpha, \beta\}\overline{x}$

22. $x = \overline{\{\alpha\}}\, x + \{\beta, \gamma\}$ 23. $x = \overline{\{\alpha, \beta\}}\, x + \{\alpha, \beta\}$

24. $x = \{\alpha\}\overline{x}$

IV. Write the following exponential formulas in the linear form:

25. a^{b^c} 26. $a + b^{c^d}$ 27. $a^{b+c} + d^e$ 28. $a^{b^{c+d}}$

29. a^{b+a^c} 30. $a^{b^{a+b^{\bar{a}}}}$ 31. $a^{ab^{\bar{c}}}$ 32. $a^{\bar{b}a^b}$

33. $a^{\bar{b}^{\bar{c}}+c^{\bar{b}}}$ 34. $a^{0^{\bar{b}+0^{\bar{a}}}}$

V. Analyze the equations given in exponential form:

35. $x = a^x$ 36. $x = x^x$ 37. $x = x^{x^x}$

38. $x = a^x + b^x$ 39. $x = a^x + b$ 40. $x = a^x + x$

41. $x = a^{b^x}$ 42. $x = (a+b)^{x+c}$ 43. $x = (ax)^{bx}$

44. $x = (a+x)^{b+x}$ 45. $x = x^{a^x}$ 46. $x = x^{a+x^b}$

VI. Analyze the equations defined on set M of all subsets of the universal set $\{\alpha, \beta, \gamma\}$:

47. $x = 0^{\{\alpha,\beta\}\bar{x}}$ 48. $x = \{\alpha,\beta\}^{\overline{\{\gamma\}+\bar{x}}}$

49. $x = \{\alpha\}^{\{\alpha,\gamma\}+\overline{\{\beta\}x}}$ 50. $x = \{\alpha,\gamma\}^{\{\beta\}+\{\alpha\}x}$

Chapter 2

I. We will designate the solid sides by R, and the broken ones by $\bar{R}$. Consider graph

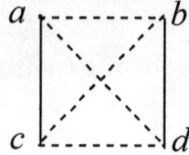

Fig. A1-1. Stratifiable graph

This graph can be stratified. It can be divided into two minimal strata on $\bar{R}$:

$$<a, c> \bar{R} <b, d>$$

In a similar way, divide the following graphs into minimal strata:

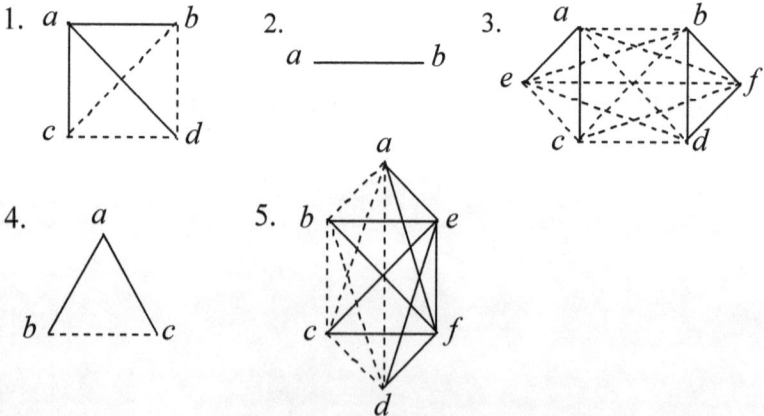

Fig. A1-2. Problems 1-5

II. For each graph, determine whether it is $S_{(4)}$:

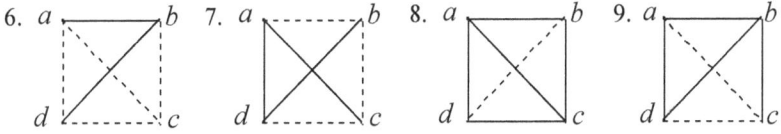

Fig. A1-3. Problems 6-9

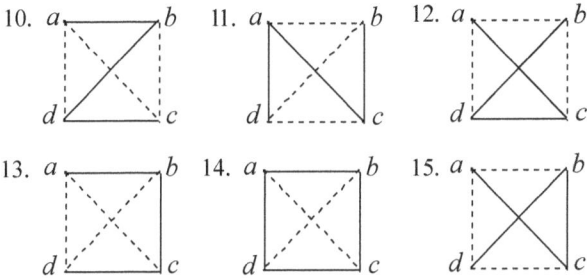

Рис. A1-4. Problems 10-15

III. Determine whether each graph contains subgraphs $S_{(4)}$, and if so, how many:

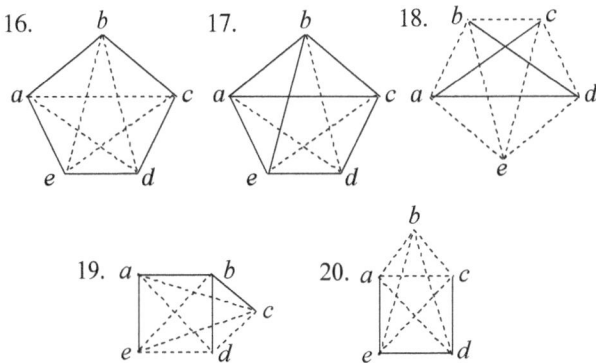

Fig. A1-5. Problems 16-20

Chapter 3

I. Determine whether each graph can be decomposed:

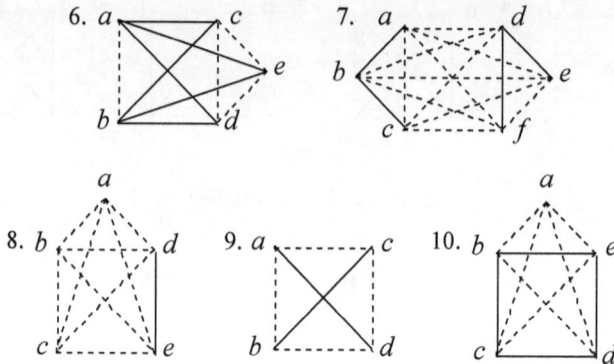

Fig. A1-6. Problems 1-5

II. Write polynomials for the following decomposable graphs:

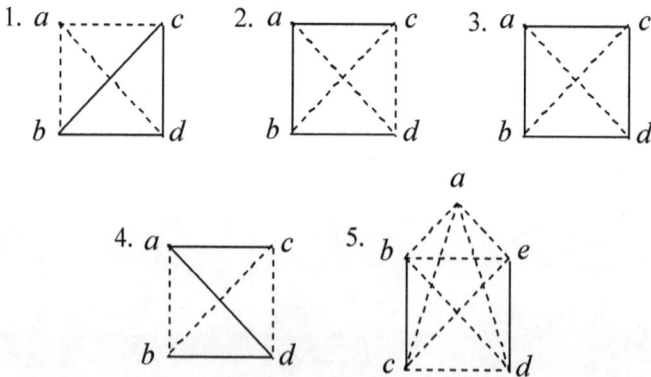

Fig. A1-7. Problems 6-10

III. Draw graphs corresponding to the following polynomials:

11. $a(b + c)$ 12. $a + bc$ 13. $a + b + c$

14. abc 15. $ab + cd$ 16. $ab + c\,(d + e)$

17. $(a + bc)\,d$ 18. $a + b + cd$ 19. $(a + b)(c + d)$

20. $(a + b)(c + de)$ 21. $(ab + c)de$

IV. Write diagonal forms for the following polynomials:

22. $a + b + c$ 23. $a + bc$ 24. $(a + b)(c + d)$

25. $a(b + c)$ 26. $d + a(b + c)$ 27. $abc\,(d + e)$

28. $a\,(b + c) + de$ 29. $(ab + cd)\,e$ 30. $ab\,(c + d)$

31. $a + bc + def$ 32. $a + b(c + d\,(e + f))$

33. $a + bc\,(d + ef)$ 34. $(a + b + c)\,(d + ef)\,g$

35. $(ab + cd)\,(eh + fg)$ 36. $abc + d\,(g + ef)$

Chapter 4

I. Taking Fig. 4.2.3 as an example, draw diagrams of partial order for the following diagonal forms

$$[c] + [d]$$
$$[a] + [b] \qquad\qquad\qquad [a]\,[b]\,[c + d]$$
1. $[a + b]$ 2. $[ab\,(c + d)]$

$$[b]\,[c]$$
$$[a] + [bc]$$
$$[a + bc] \qquad\qquad [d]$$
3. $[(a + bc)\,d]$

II. Transform each of the following diagonal forms into linear notation (remember that diagonal forms are exponential expressions in which parentheses and brackets are equivalent):

$$[b]\,[c] \qquad\qquad\qquad\qquad [b] + [c]$$
$$[a] + [bc] \qquad\qquad\qquad [a]\,[b + c]$$
4. $[a + bc]$ 5. $[a\,(b + c)]$

$$[b]\,[c]$$
$$[bc] \qquad + [d]$$
$$[a]\,[bc + d]$$
6. $[a\,(bc + d)]$

$$[c] + [d]$$
$$[b]\,[c + d]$$
$$[a] + [b\,(c + d)]$$
7. $[a + b\,(c + d)]$

III. Analyze equations of the type $a = \Phi(a)$, where $\Phi(a)$ is as follows:

$$\begin{array}{cc} [a]\,[b] & [c]\,[d] \\ [ab] & +\,[cd] \end{array}$$

8. $[ab + cd]$

$$[b] + [c] + [d] + [e]$$
$$[a]\,[b + c + d + e]$$

9. $[a\,(b + c + d + e)]$

$$\begin{array}{cc} [b]\,[c] & [e]\,[f\,] \\ [a] + [bc] & [d] + [ef\,] \\ [a + bc] & [d + ef\,)] \end{array}$$

10. $[(a + bc)\,(d + ef\,)]$

IV. The graph of relations between subjects a, b, c and d is given in Fig. A1-8:

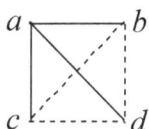

Fig. A1-8. Relations graph for problem 11

The universal set of actions is $1 = \{\alpha, \beta, \gamma, \delta\}$, and the influence matrix is as follows:

Table A1-1. Matrix of influences for problem 11

	a	b	c	d
a	a	$\{\beta\}$	$\{\alpha,\beta\}$	1
b	$\{\alpha\}$	b	$\{\alpha\}$	$\{\beta\}$
c	$\{\alpha\}$	$\{\alpha\}$	c	$\{\beta\}$
d	$\{\alpha,\beta\}$	$\{\alpha\}$	$\{\beta\}$	d

11. Find the choice of each subject

V. Consider the situation described in section 4.5, assuming that subject a_1 chooses alternative 1. The universal set is $\{\alpha, \beta, \gamma, \delta\}$.

12. Find the set of actions attractive for the group (P).

13. Find the set of actions attractive for subject a_1 (W).

14. Find the set of actions prohibited for subject a_1 $(\overline{P}W)$.

VI. For a subject with universal set $1 = \{\alpha, \beta, \gamma\}$ find the set of actions that are always chosen (R) and the set of actions that are never chosen (S), if

15. $A = \{\alpha, \beta\}, B = 0;$ 16. $A = 1, B = \{\alpha, \beta\};$

17. $A = 1, B = 0;$ 18. $A = 0, B = 0$

Chapter 5

I. Using the graph of relations in Fig. 5.1.1, find:

1. The state of subject b with the matrix of influences in Table 5.1.2

2. The states of subjects a, b, c, d, e with the following matrix of influences:

Table A1-2. Matrix of influences for problem 2

	a	b	c	d	e
a	a	1	0	0	0
b	0	b	0	0	1
c	0	1	c	1	1
d	1	1	0	d	0
e	0	0	0	0	e

II. Determine whether subjects a, b, c, d, e, corresponding to equations (5.1.2), can be all at the same time

3. in the active state

4. in the passive state

5. in a state of frustration

6. in a state of free choice

III. The graph of relations is as follows:

Appendix

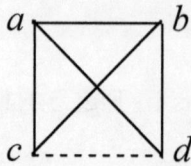

Fig. A1-9. Graph of relations for problems 7 and 8

7. Construct the matrix of influences for which subjects *a* and *b* would be in a state of frustration, and subjects *c* and *d* in a state of free choice.

8. Is it possible to change only one relation in such a way that all group members would be in the active state, independently of the matrix of influences?

Chapter 6

I. A graph of relations is as follows:

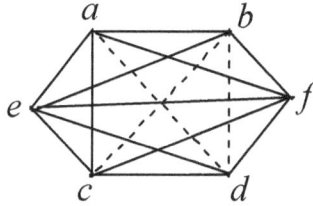

Fig. A1-10. Graph of relations for problems 1-4

Find the possible choices for subject a if the order of other subjects' significance is:

1. b, c, d, e, f 2. b, f, e, d, c

3. d, b, c, f, e 4. f, d, e, c, b

II. Consider a group with the graph of relations in Fig. A1-11:

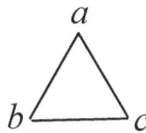

Fig. A1-11. Graph of relations for problems 5 and 6

The universal set for a is $\{\alpha\}$, for b $\{\beta, \gamma\}$, for c $\{\delta, \eta, \theta\}$.

5. What is each subject's choice?

6. What alternatives can they realize, if action β is incompatible with γ, and action δ is incompatible with η?

Chapter 7

I. The following figures show graphs or relations between subjects, but the relation between a and b is omitted. What should this relation be (cooperation or conflict) to make the group superactive?

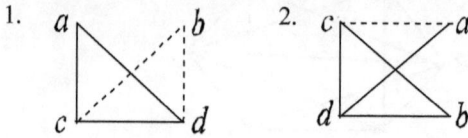

Fig. A1-12. Problems 1 and 2

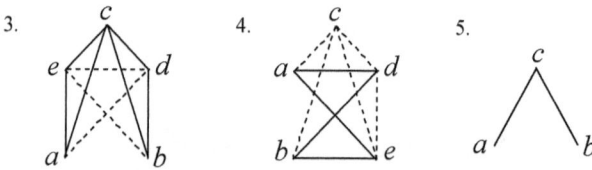

Fig. A1-13. Problems 3-5

II. What should be the relation of a and b so that the group does not become superactive?

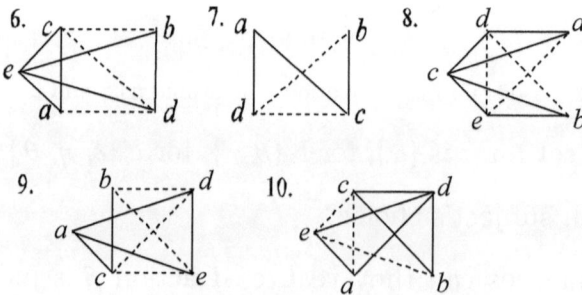

Fig. A1-14. Problems 6-10

Chapter 8

I. Consider a group with the relations given in Fig. A1-15:

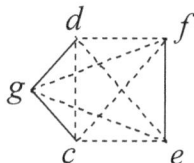

Fig. A1-15. Relations graph for problems 1-4

1. Is this group superactive?

2. Will this group be superactive if a peacemaker a joins it?

3. Will this group be superactive if two peacemakers a and b, who are in union, join the group?

4. Will this group be superactive if two peacemakers a and b, who are in conflict, join the group?

II. Consider the graph of relations

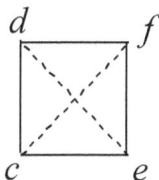

Fig. A1-16. Relation graph for problems 5-7

5. Is this group superactive?

6. Will this group be superactive if subject a, who is friendly with everyone else, joins the group?

7. Will this group be superactive if subject a, who is in conflict with everyone else, joins the group?

Chapter 9

I. Manipulation through influence

Scheme:

"a wants b to choose x; to achieve this a exerts influence x"

Consider the graph

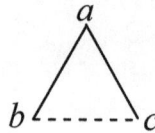

Fig. A1-17. Relations graph for problems 1-8

Can a apply the given scheme to incline b to choose 1, if

1. $c = 1$ 2. $c = 0$ 3. $d = 1$ 4. $d = 0$?

Can a apply the given scheme to incline b to choose 0, if

5. $d = 0$ 6. $c = 1$ 7. $d = 0, c = 0$ 8. $d = 0, c = 1$

Scheme (inverse control):

"a wants b to choose x, but to achieve this a has to exert influence $\bar{x}$ "

Consider the relations

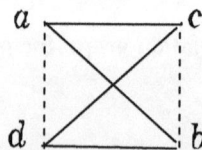

Fig. A1-18. Relations graph for problem 9

9. Let a want b to choose 1. Can he achieve this by using inverse control?

Scheme:

"*a* wants *b* to have freedom of choice and exerts influence *x* to achieve this"

Consider the graph of relations

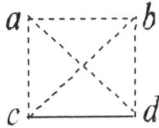

Fig. A1-19. Relations graph for problems 10-13

Can *a* move *b* into a state of free choice if

10. $c = 1, d = 1$;   11. $c = 0, d = 1$;

12. $c = 1, d = 0$;   13. $c = 0, d = 0$.

Scheme:

"*a* wants *b* to become incapable of making a choice and exerts influence *x*"

Consider the graph

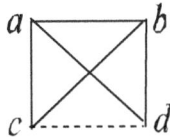

Fig. A1-20. Graph of relations for problems 14-17

Can *a* fulfill his task if

14. $c = 0, d = 0$;   15. $c = 1, d = 1$;

16. $c = 0, d = 1$;   17. $c = 1, d = 0$?

II. Manipulations with relations

Scheme:

"a wants b to have a chance to obtain the freedom of choice and changes relation (a, b)"

Let the graph of relations be as follows:

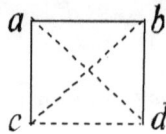

Fig. A1-21. Relations graph for problems 18 and 19

18. Will a achieve the goal by changing the relation (a, b)?

Scheme:

"a wants b to have a chance of obtaining freedom of choice and leaves the group"

19. Will a reach his goal if he leaves the group whose relations are given in Fig. A1-21?

III. Manipulation through the order of subjects' significance

Scheme:

"a does not want to be removed from b's graph of relations and changes the order of subjects' significance for b"

Consider non-decomposable graph of relations:

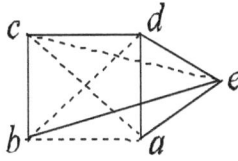

Fig. A1-22. Graph of relations for problems 20-22

Will *a* remain in the graph for the following orders of subjects' significance:

20. *c, d, a, e* ;   21. *e, c, a, d* ;   22. *d, e, a, c*

Chapter 10

I. The son corresponds to the following graph of relations:

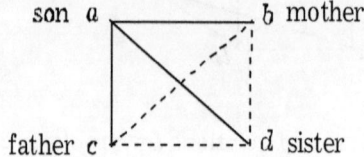

Fig. A1-23. Relations graph for problems 1-3

The son's universal set is $\{\alpha, \beta, \gamma\}$.

1. What will be the son's choice if $b = \{\alpha\}$, $c = \{\alpha\}$, $d = \{\alpha\}$?

2. What influence must the mother exert to make the son marry?

3. In what state will the son be, if mother, father, and a sister incline him not to marry?

II. Consider the graph in Fig.10.2.1. Let John and Tom be moved to another cell. Everyone else insists on his own plan of escape, and John's and Tom's plans are not considered.

4. What are choices made by Bob, Peter, and Larry?

III. Consider the scenario with the lost cigarette (Fig.10.3.1). Tom is moved to the prison hospital, but John continues to suspect him. Relations between John, Bob, and Peter remain the same. All three of them sympathize with Tom and feel solidarity with him.

5. What is John's choice?

IV. Consider the graph in Fig.10.4.1.

6. What is Edward's choice if the significance order of his group members is as follows: Alex (a), Bart (b), David (d), Gregory (g), and these influences do not change?

Chapter 11

I. Consider the scenario corresponding to Fig. 11.1.1.

1. The conflict between the political elite and the military has changed to cooperative relations and no further changes occurred. What will be the choice of the political elite?

2. The secret police are disbanded, but no other changes take place. What will be the choice of the political elite?

3. The military obtain power. The population is passive ($d = 0$). The political elite exists as an active force. What decision will be made by the military?

II. Consider the graph in Fig. 11.2.1.

4. Parties f and g are disbanded, in addition; also, $d = 0$ and $e = 0$. What choice will the President make?

5. A conflict takes place between the President and party c (Fig.11.2.1). Also, $b = \{a\}$, $c = \{a\}$, $g = 0$. What choice will be made by the President?

6. The President is overthrown. Party g takes power. There is discussion of war with a neighboring country. The set of alternatives is $\{1, 0\}$, where 1 means war and 0 means peace, $c = 0$, $b = 0$, $d = 0$, $e = 1$, $f = 1$. In what state is party g?

III. Consider the scenario corresponding to Fig. 11.3.1.

7. The undercover police agents succeed in making gang a break with gangs b and e. Will the group cease to be superactive?

Chapter 12

I. Consider the scenario for Fig. 12.1.1. Let the matrix be as follows:

Table A1-3. Matrix of influences for problems 1 and 2

		a	b	c	d
USA	a	a	1	1	0
Soviet Union	b	1	b	0	1
England	c	0	0	c	1
Germany	d	0	1	0	d

1. Suppose that the conflict between Germany and England has ended. In what state will each of the four countries be?

2. Suppose that the conflict between USA and Germany has also ended. In what state will each country be?

II. Consider the graph of relations in Fig.12.2.1.

3. I what state would each country be, if the Soviet Union and the USA were in cooperation at that time?

III. Consider the relations graph in Fig. 12.3.1.

4. Let Russia be in cooperation with Israel and the USA. In what state would each subject be?

IV. The conflict between China and Taiwan has persisted for several decades. During this time, the USA and Russia have had the greatest influence on the situation. We will consider that China and Russia are in cooperation, as are Taiwan and the USA; a conflict exists between these two pairs. Let the matrix of influences be as follows:

Table A1-4. Matrix of influences for problem 5

		a	b	c	d
China	a	a	0	0	0
Russia	b	0	b	0	0
Taiwan	c	1	0	c	1
USA	d	1	0	0	d

5. In what state is each subject?

Chapter 13

I. Consider the scenario corresponding to Fig. 13.1.1. Suppose platoon a goes over to the side of the battalion (d). The relation between a and d changes from conflict to cooperation, and the relation between a and b and c changes to conflict.

1. What will the battalion commander choose?

II. In some country there is a civil war in progress. Tribe b intends to destroy tribe d. A third tribe, c, is cooperative with b and in conflict with d. There is a peacemaker a in the country. He persuaded the leaders of the tribes c and d try to persuade tribe b to cease hostilities, by allowing b to fish in their lakes. The set of alternative for b is $\{1, 0\}$, where 1 is fighting and 0 is to stop fighting. Permission to fish in the lakes belonging to tribes c and d is their influence on tribe b, inclining b to choose stop fighting: $c = 0$, $d = 0$.

2. What else must the peacemaker do to stop the bloodshed?

III. Three cooperative countries, a, b and c have a potential enemy, country d. Countries a, b and c plan to start a common project: construction of a new military aircraft. There are two types to choose from: α and β. Plane α has better battle qualities, but is more expensive than β. Country a prefers α, and country b prefers β. Each of them tries to persuade country c to support its project. In addition, it is known that the potential enemy, country d, has begun developing project α. For country c, this is an additional argument in favor of α.

3. What plane will c support?

Chapter 14

I. Consider the trial relations represented by the graph in Fig. 14.1.2. A public prosecutor is included in the trial and is in conflict with all participants of the trial.

1. Will the trial remain perfect?

II. A public defender is included in the trial. He is conflict only with the prosecutor.

2. Will the trial remain perfect?

III. Let a conflict appear between the judge and the defendant during the trial, and let the defendant be removed from the hall.

3. Will the trial remain perfect?

Conclusion

1. Analyze the situation with two friends described in the Conclusion.

Answers and Explanations

Introduction

The anti-selfishness principle rejects actions by the subject that cause damage to the group or to society in cases when the subject is pursuing his personal goals. Thus, the pattern

advantageous for the subject – disadvantageous for the society

contradicts the anti-selfishness principle, while the patterns

advantageous for the subject – advantageous for the society,
disadvantageous for the subject – advantageous for the society,
disadvantageous for the subject – disadvantageous for the society,

do not contradict the principle.

1. Herostratus' actions follow the pattern "advantageous for the subject – disadvantageous for the society"; therefore, he does not behave in accord with the anti-selfishness principle.

2. The action of a rich man who leaves his fortune to the church in hopes to get to heaven follows the pattern "advantageous for the subject – advantageous for the society," which does not contradict the anti-selfishness principle.

3. The father's actions follow the pattern "advantageous for the subject – disadvantageous for the society." They contradict the principle.

4. The king's actions follow the pattern "disadvantageous for the subject – disadvantageous for the society." His actions do not contradict the anti-selfishness principle.

5. Danko's act personifies the pattern "disadvantageous for the subject – advantageous for the society" that corresponds to the anti-selfishness principle.

6. By organizing circuses, the ruler acts by the pattern "advantageous for the subject – advantageous for the society," which does not contradict the anti-selfishness principle.

7. The emperor acts on the pattern "advantageous for the subject – disadvantageous for the society," which contradicts the anti-selfishness principle. He saves a human life, acting as preferred by the self, but against the people's wishes.

8. Saving the human life and explaining the motives for his action, the emperor changes the preferences of the society. So, his action falls under the pattern "advantageous for the subject – advantageous for the society," i.e., it does not contradict the anti-selfishness principle.

Appendix

Chapter 1

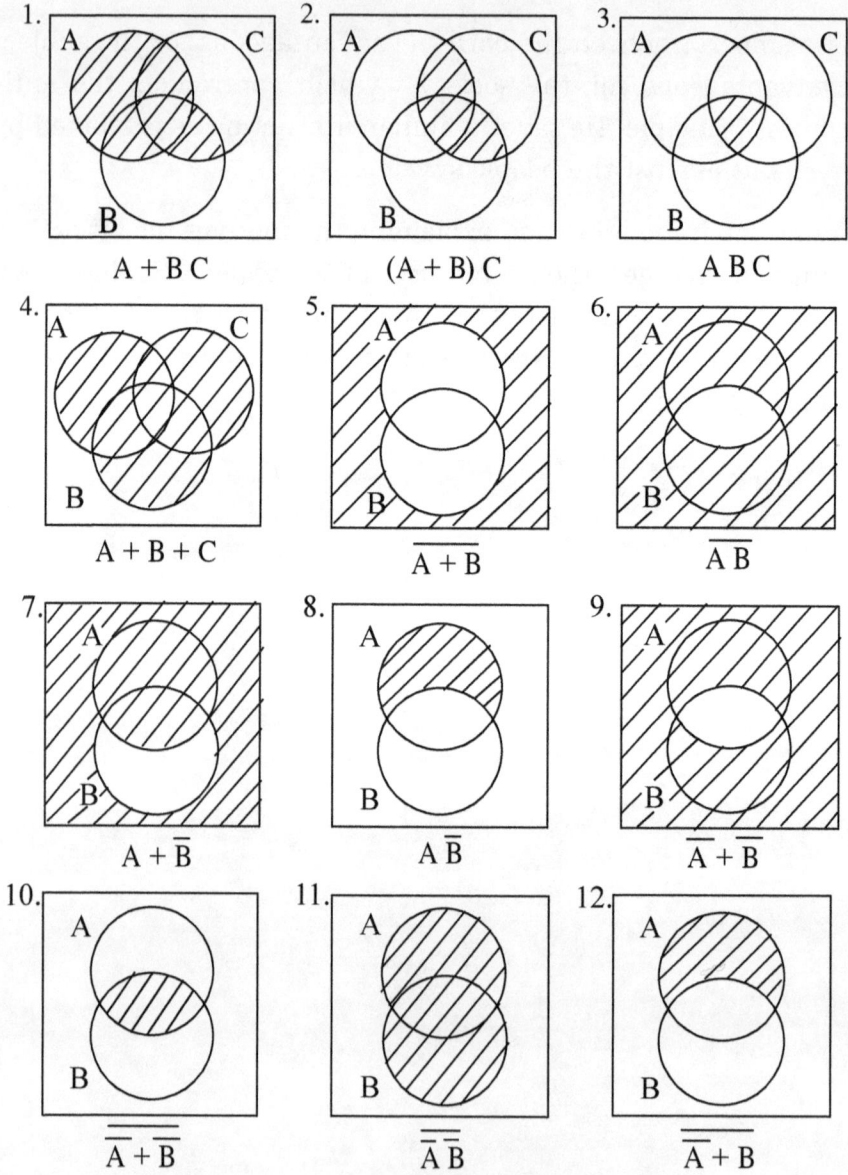

Fig. A2-1. Answers to problems 1-12

13. $(a + b) \supseteq x \supseteq b$ 14. no solution

15. $1 \supseteq x \supseteq (a + b)$ 16. $1 \supseteq x \supseteq b$

17. solution only at $c = 1$: $x = 1$

18. $\{\alpha, \beta\} \supseteq x \supseteq \{\alpha\}$ 19. no solution

20. $x = \{\alpha\}$

21. the equation can be reduced to $x = \{\beta, \gamma\}x + (\alpha, \beta\}\bar{x}$; the inequality $\{\beta, \gamma\} \supseteq \{\alpha, \beta\}$ does not hold, since the right set is not a subset of the left set; therefore, the equation has no solution

22. $x = \{\beta, \gamma\}$ 23. $1 \supseteq x \supseteq \{\alpha, \beta\}$

24. no solution   25. $a + \bar{b}c$

26. $a + b + \bar{c}d$ 27. $a + \bar{b}\bar{c} + d + \bar{e}$

28. $a + \bar{b}(c + d)$ 29. $a + \bar{b}c$

30. a 31. 1

32. a 33. a 34. $a + \bar{b}$

35. solution only at $a = 1$: $x = 1$

36. $x = 1$ 37. $1 \supseteq x \supseteq 0$

38. solution exists only if $a + b = 1$: $x = 1$

39. solution exists only if $a + b = 1$: $x = 1$

40. $x = 1$ 41. $(a + \bar{b}) \supseteq x \supseteq a$

42. solution exists only if $c \supseteq a\bar{b}$: $x = a + b$

43. solution exists only if $a + \bar{b} = 1$: $x = 1$

44. $1 \supseteq x \supseteq (a + \bar{b})$ 45. $1 \supseteq x \supseteq 0$

46. $1 \supseteq x \supseteq \bar{a}b$ 47. $1 \supseteq x \supseteq \{\gamma\}$

48. $1 \supseteq x \supseteq \{\alpha, \beta\}$ 49. $x = \{\alpha, \beta\}$

50. $x = \{\alpha, \gamma)$

Appendix

Chapter 2

1. $<a> R <b, c, d>$

2. $<a> R <b>$

3. $<a, c, e> \overline{R} <b, d, f>$

4. $<a> R <b, c>$

5. $<a, b, c, d> R <e> R <f>$

6. not $S_{(4)}$

7. $S_{(4)}$

8. not $S_{(4)}$

9. not $S_{(4)}$

10. $S_{(4)}$

11. $S_{(4)}$

12. $S_{(4)}$

13. $S_{(4)}$

14. not $S_{(4)}$

15. $S_{(4)}$

16. 5 subsets $S_{(4)}$

17. no $S_{(4)}$

18. 1 subset $S_{(4)}$

19. 2 subsets $S_{(4)}$

20. 1 subset $S_{(4)}$

Chapter 3

1. decomposable

2. not decomposable

3. decomposable

4. not decomposable

5. decomposable

6. $(a + b)(c + d + e)$

7. $b(a + c) + d(e + f)$

8. $a + b + c + de$

9. $ad + bc$

10. $a + (b + d)(c + e)$

11.

12.

13.

14.

15.

16.

17.

18.

19.

20.

21.

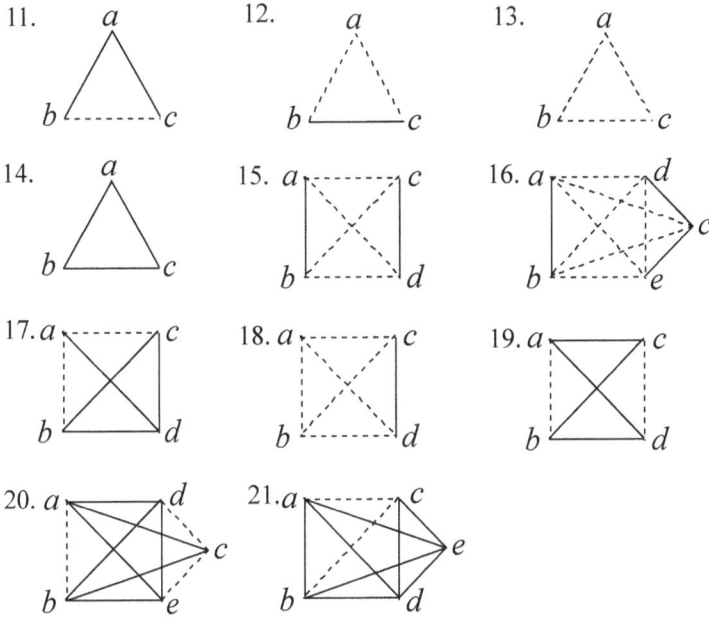

Fig. A2-2. Answers to problems 11-21

22. $[a + b + c]$
$[a] + [b] + [c]$

23. $[a + bc]$
$[a] + [bc]$
$[b][c]$

24. $[(a + b)(c + d)]$
$[a + b]$
$[c + d]$
$[a] + [b]$
$[c] + [d]$

Appendix

$$[b] + [c]$$
$$[a]\ [b + c]$$
25. $[a\ (b + c)]$

$$[b] + [c]$$
$$[a]\ [b + c]$$
$$[d] + [a\ (b + c)]$$
26. $[d + a(b + c)]$

$$[d] + [e]$$
$$[a]\ [b]\ [c]\ [d + e]$$
27. $[abc\ (d + e)]$

$$[b] + [c]$$
$$[a]\ [b + c] \qquad\qquad [d]\ [e]$$
$$[a\ (b + c)] \qquad\qquad + [de]$$
28. $[a\ (b + c) + de]$

$$[a]\ [b] \qquad\quad [c]\ [d]$$
$$[ab] \qquad + [cd]$$
$$[ab + cd] \qquad\qquad\quad [e]$$
29. $[(ab + cd)\ e]$

$$[c] + [d]$$
$$[a]\ [b]\ [c + d]$$
30. $[ab\ (c + d)]$

$$[b]\ [c] \qquad\quad [d]\ [e]\ [f]$$
$$[a] + [bc] \qquad + [def]$$
31. $[a + bc + def]$

$$[e] + [f]$$
$$[d]\ [e + f]$$
$$[c] + [d\ (e + f)]$$
$$[b]\ [c + d\ (e + f)]$$
$$[a] + [b\ (c + d\ (e + f)]$$
32. $[a + b\ (c + d\ (e + f))]$

$$[e]\ [f]$$
$$[d] + [ef]$$
$$[b]\ [c]\ [(d + ef)]$$
$$[a] + [bc\ (d + ef)]$$
33. $[a + bc\ (d + ef)]$

$$[e]\ [f]$$
$$[a] + [b] + [c] \qquad [d] + [ef]$$
$$[a + b + c] \qquad\qquad [d + ef] \qquad\qquad [g]$$
34. $[(a + b + c)\ (d + ef)\ g]$

$$[a]\ [b] \quad [c]\ [d] \qquad [e]\ [h] \qquad [f]\ [g]$$
$$[ab] \ +\ [cd] \qquad\quad [eh] \qquad +\ [fg]$$
$$[ab + cd] \qquad\qquad\qquad [eh + fg]$$
35. $[(ab + cd)\ (eh + fg)]$

$$[e]\ [f]$$
$$[g] + [ef]$$
$$[a]\ [b]\ [c] \qquad\qquad [d]\ [g + ef]$$
$$[abc] \qquad\quad +\ [d(g + ef)]$$
36. $[abc + d\ (g + ef)]$

Appendix

Chapter 4

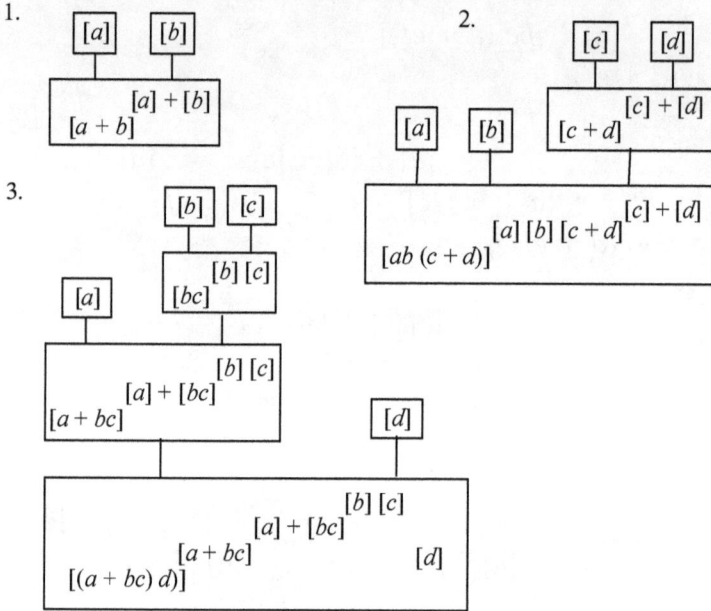

Fig. A2-3. Answers to problems 1-3

4. $a + bc$ 5. $\bar{a} + b + c$ 6. 1

7. $a + b$ 8. $(b + cd) \supseteq a \supseteq cd$

9. solution exists only if $b + c + d + e = 1$: $a = 1$

10. $a = 1$

11. Subject a is in frustration; $b : 1 \supseteq b \supseteq \{\alpha, \gamma, \delta\}$; $c : c = 1$;

$d : 1 \supseteq d \supseteq \{\beta\}$

12. $W = \{\alpha, \beta, \gamma\}$ 13. $P = \{\alpha, \beta, \gamma\}$ 14. $\bar{P}_W = \{\}$

15. $R = 0$, i.e., the set of actions always chosen is empty; $S = \{\gamma\}$

16. $R = \{\alpha, \beta\}$, $S = 0$, i.e., the set of actions never chosen is empty

17. $R = 0$, $S = 0$; both sets are empty, i.e., there are no actions always chosen, nor are there actions never chosen.

18. $R = 0$, no actions always chosen; $S = 1$, all actions are never chosen.

Appendix

Chapter 5

1. Subject b is in the active state ($b = 1$).

2. Subjects a and b are in the active state, c is in a state of frustration, and d and e are in the state of free choice.

3. They can.

For example, consider the following table:

Table A2-1. Matrix of influences for problem 3

	a	b	c	d	e
a	a	0	0	0	0
b	0	b	0	0	0
c	0	0	c	0	0
d	0	0	0	d	1
e	1	1	1	1	e

With these influences, all solutions to equation (5.1.2) are equal to 1.

4. They cannot.

To prove this, it is enough to analyze one subject e. He corresponds to the equation

$$e = Ae + B\bar{e},$$

where $A = 1$ and $B = d + \bar{c}(\bar{a} + \bar{b})$. Solutions of this equation belong to the interval

$$1 \supseteq e \supseteq B,$$

where B either 1 or 0. In the first case, subject e will be in the active state, and in the second case, in a state of free choice. Thus, no matter what the influences are, the subject will not be in the passive state.

5. They cannot.

The state of frustration appears when in the subject's equation of choice, relation $A \supseteq B$, does not hold. When solving the previous problem for subject e, we found that $A \supseteq B$ regardless of influences. Therefore, e cannot be in a state of frustration.

6. They cannot.

The state of free choice appears if the equation's solutions belong to the interval $1 \supseteq x \supseteq 0$, that is, the relation $A \supset B$ must hold. Consider the equation for subject c:

$$c = Ac + B\bar{c} \ ,$$

where $A = e + d$ and $B = e + d + \bar{a} + \bar{b}$. We see that $A \subseteq B$. Thus, subject c cannot be in a state of free choice.

7. The graph corresponds to the polynomial

$$ab \, (c + d),$$

the diagonal form

$$[c] + [d]$$
$$[a] \, [b] \, [c + d]$$
$$[ab \, (c + d)]$$

and the equations

$$a = c + d + \bar{a} + \bar{b}$$
$$b = c + d + \bar{a} + \bar{b}$$
$$c = c + d + \bar{a} + \bar{b}$$
$$d = c + d + \bar{a} + \bar{b}$$

Appendix

The following matrix moves subjects to the indicated states:

Table A2-2. Matrix of influences for problem 7

	a	b	c	d
a	a	1	1	1
b	1	b	1	1
c	0	0	c	0
d	0	0	0	d

8. It is possible.

For example, by changing the relation between c and d from conflict to cooperation.

Chapter 6

1. The choice is possible only if $b + c = 1$: $a = 1$.

The relations graph is not decomposable. After the removal of subjects f and e, the graph remains not decomposable. The removal of subject d makes it a decomposable graph:

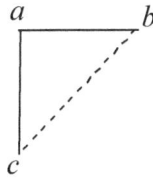

Fig. A2-4. Graph of relations for problem 1

This graph corresponds to the polynomial

$$a(b + c),$$

and the diagonal form

$$[b] + [c]$$
$$[a][b + c]$$
$$[a(b + c)]$$

and the equation for a is
$$a = (b + c)\, a + \bar{a},$$

which leads to the conclusion that choice is possible only at $b + c = 1$. So, $a = 1$.

2. $a = 1$.

3. $(b + d) \supseteq a \supseteq d$

4. $1 \supseteq a \supseteq (d + \bar{e} + \bar{f} + \bar{c})$

5. The graph corresponds to the diagonal form

Appendix

$$\frac{[a]\ [b]\ [c]}{[abc]} \equiv 1,$$

thus, every subject chooses alternative 1. The meaning of 1, however, is different for each subject:

for a, $1 = \{\alpha\}$
for b, $1 = \{\beta, \gamma\}$
for c, $1 = \{\delta, \eta, \theta\}$

6. After choosing 1, subject a can realize alternative $\{\alpha\}$; subject b either $\{b\}$ or $\{\gamma\}$; subject c can realize any of the alternatives $\{\delta\}$, $\{\eta\}$, $\{\theta\}$, $\{\delta, \theta\}$, $\{\eta, \theta\}$.

Chapter 7

1. Cooperation

2. Conflict

3. Conflict

4. Conflict

5. Cooperation

6. Cooperation

7. Conflict

8. Cooperation

9. Conflict

10. Cooperation

Appendix

Chapter 8

1. No, it is not.

The group corresponds to the polynomial

$$g(d + c) + fe$$

and the diagonal form

$$[d] + [c]$$
$$[g][d + c] \qquad\qquad [f][e]$$
$$[g(d + c)] \qquad\qquad + [fe]$$
$$[g(d + c) + fe]$$

This form is equivalent to the initial polynomial, which is equal to 0 at $g = 0$ and $f = 0$.

2. Yes, it will.

To obtain the polynomial describing the group after peacemaker a joins it, we have to multiply a by the polynomial describing the group before the arrival of the peacemaker:

$$a(g(d + c) + fe).$$

This polynomial corresponds to the diagonal form

$$[d] + [c]$$
$$[g][d + c] \qquad\qquad [f][e]$$
$$[g(d + c)] \qquad\qquad + [fe]$$
$$[a][g(d + c) + fe]$$
$$[a(g(d + c) + fe)] \qquad\qquad\qquad \equiv 1.$$

3. It will be superactive.

After two cooperating peacemakers join the group, the group corresponds to the polynomial

$$ab\,(g\,(d + c) + fe)$$

and the diagonal form

$$[d] + [c]$$
$$[g]\,[d + c] \qquad\qquad [f]\,[e]$$
$$[g\,(d + c)] \qquad\qquad + [fe]$$
$$[a]\,[b]\,[g\,(d + c) + fe)]$$
$$[ab\,(g\,(d + c) + fe)] \qquad\qquad\qquad \equiv 1.$$

4. No, it will not.

After two conflicting peacemakers join the group, the polynomial becomes

$$(a + b)\,(g\,(d + c) + fe)$$

and the diagonal form

$$[d] + [c]$$
$$[g]\,[d + c] \qquad [f]\,[e]$$
$$[a] + [b] \qquad\qquad [g\,(d + c)] \qquad + [fe]$$
$$[a + b] \qquad [g\,(d + c) + fe]$$
$$[(a + b)\,(g\,(d + c) + fe)] \qquad\qquad\qquad ,$$

which is equivalent to

$$a + b + (\overline{g} + \overline{dc})(\overline{f} + \overline{e}).$$

At $a = 0$, $b = 0$, $f = 1$ and $e = 1$ this expression is equal to 0.

5. The group is not superactive.

Its polynomial is

$$(d + e)\,(c + f)$$

and its diagonal form is

Appendix

$$[d] + [e] \qquad [c] + [f]$$
$$[d + e] \qquad [c + f]$$
$$[(d + e)\,(c + f)] \qquad\qquad ,$$

which is equivalent to the initial polynomial equal to zero when $d = 0$ and $e = 0$.

6. The group will not be superactive.

After subject a joins the group, its polynomial becomes

$$a\,(d + e)\,(c + f)$$

and its diagonal form is

$$[d] + [e] \qquad [c] + [f]$$
$$[a]\,[d + e] \qquad [c + f]$$
$$[a(d + e)\,(c + f)] \qquad\qquad ,$$

which is equivalent to

$$(d + e)\,(c + f) + \bar{a}\,.$$

This expression is equal to zero at $a = 1$, $d = 0$, $e = 0$.

7. It will.

The group corresponds to the polynomial

$$a + (d + e)\,(c + f)$$

and the diagonal form

$$[d] + [e] \qquad [c] + [f]$$
$$[d + e] \qquad [c + f]$$
$$[a] + [(d + e)\,(c + f)]$$
$$[a + (d + e)\,(c + f)] \qquad\qquad \equiv 1.$$

Chapter 9

1. Yes, it can.

The graph corresponds to the polynomial

$$(a + d)(c + b),$$

and the diagonal form

$$[a] + [d] \qquad [c] + [b]$$
$$[a + d] \qquad [c + b)]$$
$$[(a + d)(c + b)]$$

and the equation for b is

$$b = (a + d)(c + b).$$

When $c = 1$, subject a with influence $a = 1$ makes subject b choose alternative 1.

2. Cannot   3. Cannot   4. Cannot   5. Can
6. Cannot   7. Can   8. Can   9. Can

The equation for b is
$$b = a(b + c) + \bar{a}.$$

To make b to choose 1, a must exert influence 0.

10. Cannot.

The graph corresponds to the polynomial

$$a + b + cd,$$

and the diagonal form

$$[c][d]$$
$$[a] + [b] + [cd]$$
$$[a + b + cd]$$

and the equation for b is

Appendix

$$b = a + b + cd .$$

When $c = 1$ and $d = 1$, subject b chooses 1, independently of a's influence.

11. Can.   12. Can.   13. Can.   14. Can.

The graph corresponds to the polynomial

$$ab(c + d),$$

and the diagonal form

$$[c] + [d]$$
$$[a]\,[b]\,[c + d]$$
$$[ab(c + d)] \qquad ,$$

and the equation for b is

$$b = c + d + \bar{a} + \bar{b} .$$

When $c = 0$ and $d = 0$, a must exert influence $a = 1$, so that b becomes incapable of making a choice.

15. Cannot.   16. Cannot.   17. Cannot.   18. He will.

After the relation (a, b) has changed, the graph corresponds to the polynomial

$$ac + bd,$$

and the diagonal form

$$[a]\,[c] \qquad [b]\,[d]$$
$$[ac] \qquad + [bd]$$
$$[ac + bd]$$

and the equation for b is

$$b = ac + bd.$$

Subject b has a chance to obtain freedom of choice. This will happen when $c = 0$ and $d = 1$.

19. He will.

20. Will not.

After subject e is removed, the graph remains not decomposable, and the next one to be removed is subject a.

21. Will not.

22. Will remain.

Chapter 10

1. The son cannot make a choice.

He corresponds to the polynomial

$$a (b + c + d),$$

and to the diagonal form

$$[b] + [c] + [d]$$
$$[a] [b + c + d]$$
$$[a (b + c + d)]$$

,

and the equation is

$$a = (b + c + d)a + \bar{a} .$$

The condition for this equation to have a solution is $b + c + d = \{\alpha, \beta, \gamma\} = 1$. In the given case, $b + c + d = \{\alpha\} \subset 1$; thus, choice is impossible.

2. $b = (\alpha, \beta, \gamma\} = 1$

3. The son will be in a state of frustration.

4. Bob (a_3): $1 \supseteq a_3 \supseteq \{\alpha_3, \alpha_4\}$
   Peter (a_4): $1 \supseteq a_4 \supseteq \{\alpha_3, \alpha_4\}$
   Larry (a_5): cannot make a choice

After John and Tom are moved to another cell, the graph of relations becomes

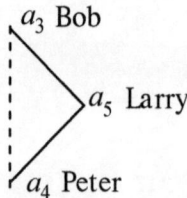

a_3 Bob

a_5 Larry

a_4 Peter

Fig. A2-5. Graph of relations for problem 4

The graph corresponds to the polynomial

$$a_5 (a_3 + a_4),$$

the diagonal form

$$[a_3] + [a_4]$$
$$[a_5] [a_3 + a_4]$$
$$[a_5 (a_3 + a_4)]$$

and the equations

$$a_3 = a_5(a_3 + a_4) + \bar{a}_5$$
$$a_4 = a_5(a_3 + a_4) + \bar{a}_5$$
$$a_5 = a_5(a_3 + a_4) + \bar{a}_5$$

The matrix of influences is

Table A2-3. Matrix of influences for problem 4

	a_3	a_4	a_5
a_3	a_3	$\{\alpha_3\}$	$\{\alpha_3\}$
a_4	$\{\alpha_4\}$	a_4	$\{\alpha_4\}$
a_5	$\{\alpha_5\}$	$\{\alpha_5\}$	a_5

By solving the equations with the variables' values given in the matrix, we obtain the choices indicated above.

5. John chooses alternative 1.

After Tom is sent to the prison hospital, the graph becomes

Appendix

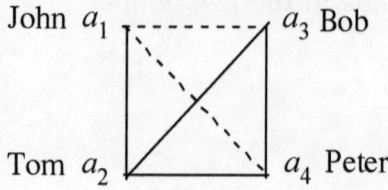

Fig. A2-6. Relations graph for problem 5

The graph corresponds to the polynomial

$$a_2(a_1 + a_3\, a_4)$$

and the diagonal form

$$[a_3]\,[a_4]$$
$$[a_1] + [a_3\, a_4]$$
$$[a_2]\,[a_1 + a_3\, a_4]$$
$$[a_2(a_1 + a_3\, a_4)] \qquad\qquad \equiv 1.$$

Thus, John chooses $\{a_2,\ a_3,\ a_4\} = 1$. This choice can be interpreted as John's conclusion that the cigarette was not lost but stolen by someone in the cell.

6. Edward cannot make a decision; he is in a state of frustration.

After Gregory's removal, the graph remains not decomposable. In accord with his order of significance, Edward removes David from consideration and the graph becomes

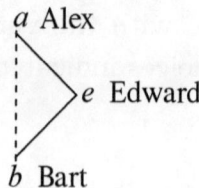

Fig. A2-7. Relations graph for problem 6

It corresponds to the polynomial

$$e\,(a + b),$$

and the diagonal form

$$[a] + [b]$$
$$[e]\,[a + b]$$
$$[e\,(a + b)]$$

and the equation for Edward is

$$e = (a+b)e + \bar{e}\,,$$

$a = \{\delta\}$, $b = \{\delta\}$, so $(a + b) \subset 1$. The equation has no solution.

Appendix

Chapter 11

1. The choice of the political elite belongs to interval $1 \subseteq a \subseteq \{\beta\}$, i.e., it chooses one of two alternatives: $1 = \{\alpha, \beta\}$, $\{\beta\}$. Choosing 1 is interpreted as the decision not to be inactive but choose some economic policy, choosing $\{\beta\}$ as the decision to have a market economy. The situation corresponds to the graph

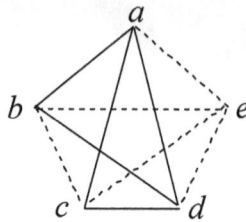

Fig. A2-8. Graph of relations to problem 1

the polynomial

$$e + ad\,(b + c),$$

and the diagonal form

$$[b] + [c]$$
$$[a]\,[d]\,[b + c]$$
$$[e] + [ad\,(b + c)]$$
$$[e + ad\,(b + c)]$$

and the equation for a is

$$a = (d + e)a + e\bar{a}\,.$$

Note that, in this case, the choice of the political elite depends on the positions of the population(d) and of business (e): $d = \{\alpha\}$, $e = \{\beta\}$. With these values

$$1 \supseteq a \supseteq \{\beta\}\,.$$

2. $a = 1$. The elite will not be inactive.
Here is the graph of the situation:

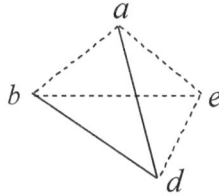

Fig. A2-9. Graph of relations for problem 2

the polynomial
$$e + d\,(a + b),$$

the diagonal form
$$[a] + [b]$$
$$[d]\,[a + b]$$
$$[e] + [d\,(a + b)]$$
$$[e + d\,(a + b)]$$

and the equation for a
$$a = e + d\ .$$

In this case, the choice of the political elite does not depend on military influence. For $d = \{a\}$,
$e = \{\beta\}$

$$a = 1\ .$$

The political elite chooses an active line of behavior and will decide for itself what economic course to take.

3. The military will make the decision that business dictates.
The equation for the military is

$$b = e + d\,(b + ac),$$

which differs from (11.1.3) only by the variable b on the left side.

Appendix

Since $d = 0$,

$$b = e .$$

4. The president has freedom of choice.
The following graph corresponds to the situation:

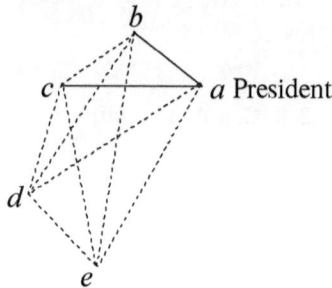

Fig. A2-10. Graph of relations for problem 4

the polynomial

$$d + e + a (b + c),$$

the diagonal form

$$[b] + [c]$$
$$[a] [b + c]$$
$$[d] + [e] + [a (b + c)]$$
$$[d + e + a (b + c)]$$

and the equation for a

$$a = a + d + e.$$

We substitute the values of influences $d = 0$, $e = 0$ and obtain:

$$a = a.$$

5. The president chooses candidate α.
The situation corresponds to the graph

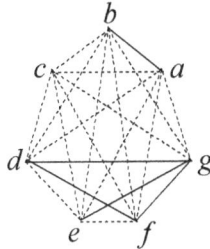

Fig. A2-11. Graph of relations for problem 5

the polynomial

$$ab + c + g\,(e + df),$$

and the diagonal form

$$[d]\,[f]$$
$$[e] + [df]$$
$$[a]\,[b] \qquad\qquad [g]\,[e + df]$$
$$[ab] \qquad + [c] + [g\,(e + df)]$$
$$[ab + c + g\,(e + df)]$$

and the equation for a is

$$a = ab + c + g\,(e + df).$$

After substitution we find

$$a = \{a\}.$$

6. Party g is in a state of free choice.

The situation corresponds to the graph

Appendix

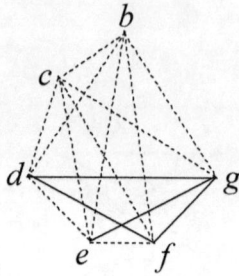

Fig. A2-12. Graph of relations for problem 6

the polynomial

$$b + c + g\,(e + df),$$

and the diagonal form

$$\begin{aligned}
& & & [d]\,[f] \\
& & & [e] + [df] \\
& & [g]\,[e + df] & \\
& [b] + [c] + [g\,(e + df)] & & \\
[b + c + g\,(e + df)] & & &
\end{aligned}$$

and the equation for g is

$$g = b + c + g\,(e + df).$$

We substitute the values of the variables and obtain

$$g = g,$$

i.e., party g has freedom of choice.

7. The group of gangs ceases to be superactive.
The situation corresponds to the graph

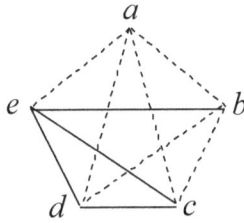

Fig. A2-13. Graph of relations for problem 7

the polynomial

$$a + e\,(b + c\,d),$$

and the diagonal form

$$[c]\,[d]$$
$$[b] + [c\,d]$$
$$[e]\,[b + c\,d]$$
$$[a] + [e\,(b + c\,d)]$$
$$[a + e\,(b + c\,d)] \qquad\qquad ,$$

which is equivalent to the initial polynomial not equal identically to 1. For example, if all variables take on the value of zero, the polynomial will be equal to zero.

Appendix

Chapter 12

1. All countries are in a state of free choice.

After relations change between England and Germany, the graph becomes as follows:

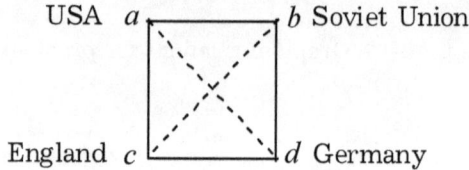

Fig. A2-14. Relations graph for problem 1

the polynomial:

$$(a + d)(b + c),$$

the diagonal form:

$$[a] + [d] \qquad [b] + [c]$$
$$[a + d] \qquad [b + c]$$
$$[(a + d)(b + c)]$$

and the equations of the type

$$x = (a + d)(b + c),$$

where $x = a, b, c, d$.

We substitute the variables' values from the matrix of influences and find that each country is in a state of free choice.

2. Soviet Union is in a state of free choice, and other countries are in the active state.

The graph of relations is as follows:

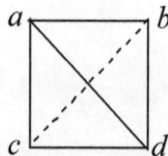

Fig. A2-15. Relations graph for problem 2

the polynomial:

$$ad (b + c),$$

the diagonal form:

$$[b] + [c]$$
$$[a] [d] [b + c]$$
$$[ad (b + c)]$$

and the equations for the countries:

$$x = b + c + \bar{a} + \bar{d} \; ,$$

where $x = a, b, c, d$.

Substituting the variables' values from the influences matrix, we obtain the states of the countries.

3. The subjects are in the superactive state.
The graph:

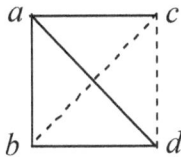

Fig. A2-16. Relations graph for problem 3

the polynomial:

$$a (c + bd),$$

and the diagonal form:

$$[b] [d]$$
$$[c] + [bd]$$
$$[a] [c + bd]$$
$$[a (c + bd)] \qquad \equiv 1.$$

Appendix

4. The subjects are in the superactive state.
The graph of the situation:

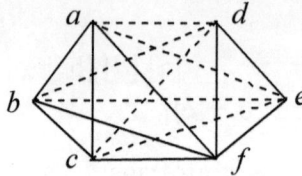

Fig. A2-17. Graph of relations for problem 4

the polynomial:

$$f(abc + de)$$

and the diagonal form:

$$
\begin{array}{ccc}
 & [a]\,[b]\,[c] & [d]\,[e] \\
 & [abc] & + [de] \\
[f]\,[abc + de] & & \\
[f(abc + de)] & & \equiv 1.
\end{array}
$$

5. China is in the active state, Russia and Taiwan are in the passive state, USA is in a state of free choice.
The graph:

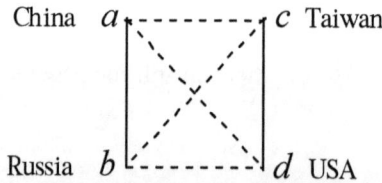

Fig. A2-18. Graph of relations for problem 5

the polynomial:

$$ab + cd,$$

the diagonal form:

$$
\begin{array}{cc}
[a]\,[b] & [c]\,[d] \\
[ab] & + [cd] \\
[ab + cd] &
\end{array}
$$

and the equations:

$$x = ab + cd,$$

where $x = a, b, c, d$. We substitute the values from the matrix of influences and find the states of the countries.

Appendix

Chapter 13.

1. The battalion commander chooses crossing A.
The graph after relations change is

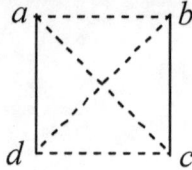

Fig. A2-19. Graph of relations for problem 1

the polynomial:

$$ad + bc \,,$$

the diagonal form

$$\begin{array}{cc} [a]\,[d] & [b]\,[c] \\ [ad] & + [bc] \\ [ad + bc] \end{array}$$

and the equation for the battalion commander:

$$d = ad + bc \,.$$

Crossing A attracts the battalion commander because it has not been captured by the enemy. This means that the commander is under influence $a = \{\alpha\}$. Crossings b and c repel the commander:

$$b = \overline{\{\beta\}} = \{\alpha, \gamma\},$$
$$c = \overline{\{\gamma\}} = \{\alpha, \beta\}.$$

By substituting these values into the equation, we find that the battalion commander chooses $\{\alpha\}$, i.e., crossing A, which is in the hands of a platoon that has come over to the side of the battalion.

2. The peacemaker must cease his activity.
The situation corresponds to the graph

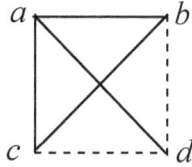

Fig. A2-20. Graph of relations for problem 2

the polynomial

$$a\,(bc + d)$$

and the diagonal form

$$
\begin{array}{ll}
& [b]\,[c] \\
& [bc] \qquad + [d] \\
& [a]\,[bc + d] \\
[a\,(bc + d)] & \qquad\qquad \equiv 1.
\end{array}
$$

The group is in the superactive state. Thus, tribe b continues to destroy tribe d. Although the peacemaker persuades the leaders of the tribes c and d to incline tribe b to stop the bloodshed, this does not change the situation. Subject b will be in the active state independently of any influences from subjects c and d.

If the peacemaker stops his activity, however, the problem will be solved. The graph will become:

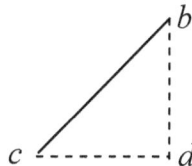

Fig. A2-21. Graph of relations for problem 2 after the peacemaker leaves.

the polynomial

$$d + bc,$$

and the diagonal form

Appendix

$$[b]\ [c]$$
$$[d] + [bc]$$
$$[d + bc]$$

and the equation for b is

$$b = d + bc.$$

The influence on b to cease hostilities is $c = 0$, $d = 0$. Thus, $b = 0$, which means that the bloodshed stops.

3. Country c prefers plane α.
The situation corresponds to the graph

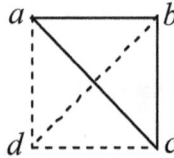

Fig. A2-22. Relations graph for problem 3

the polynomial

$$d + abc,$$

and the diagonal form

$$[a]\ [b]\ [c]$$
$$[d] + [abc]$$
$$[d + abc]$$

and the equation for c is

$$c = d + abc.$$

Country a inclines country c to choose project α: $a = \{\alpha\}$; country b inclines country c to choose project β: $b = \{\beta\}$. The potential enemy, d, by its example inclines c to choose α: $d = \{\alpha\}$. By substituting these values into the equation for c, we find that
$c = \{\alpha\}.$

Chapter 14

1. The trial will no longer be perfect.
After the public prosecutor[1] joins the judicial process, the graph becomes:

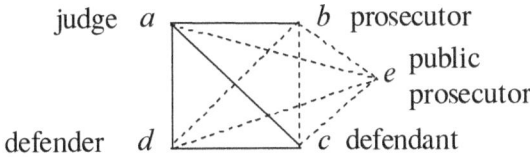

judge *a*   *b* prosecutor

e public
prosecutor

defender *d*   *c* defendant

Fig. A2-23. Graph of relations for problem 1

This graph corresponds to the polynomial

$$e + a (b + cd),$$

and the diagonal form

$$
\begin{array}{l}
\qquad\qquad\qquad [c] [d] \\
\qquad\qquad [b] + [cd] \\
\qquad [a] [b + cd)] \\
[e] \ + [a (b + cd)] \\
[e + a (b + cd)]
\end{array}
$$

and the equation for the judge is:

$$a = e + a (b + cd).$$

Suppose the defender demands that the case be dropped ($d = 0$). After that, the equation for the judge becomes

$$a = e + ab.$$

We see that the judge depends on the two prosecutors b and e. If they demand conviction ($b = e = \{\alpha\}$), the defendant will be convicted; if they demand acquittal ($b = e = \{\beta\}$), the defendant will be acquitted.

---

[1]In the Soviet Union, so-called "public representatives" were included in the process of a trial as public prosecutors or public defenders.

Appendix

Finally, if they demand that the case be dropped ($b = e = 0$), the case will be dropped. This trial is not perfect, because there are cases in which the judge chooses alternatives that differ from 1.

2. The trial remains perfect.
After the public defender joins the judicial process, the graph becomes

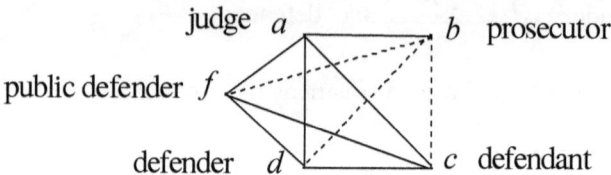

Fig. A2-24. Relations graph for problem 2

the polynomial

$$a\,(b + cdf),$$

and the diagonal form

$$[a\,(b + cdf)] \quad [a]\,[b + cdf\,] \quad [b] + [cdf\,] \quad [c]\,[d]\,[f\,] \equiv 1.$$

Therefore, regardless of influences, the judge chooses alternative 1.

3. The trial will not be perfect.
After the defendant is removed from the hall, the graph becomes

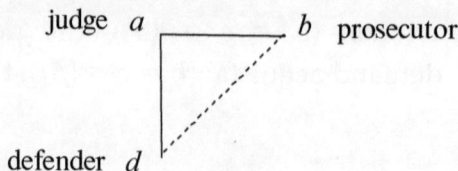

Fig. A2-25. Relations graph for problem 3

the polynomial

$$a\,(b + d),$$

and the diagonal form

$$[b] + [d]$$
$$[a]\,[b + d]$$
$$[a\,(b + d)]$$

and the equation for the judge is

$$a = (b + d)a + \bar{a}\,.$$

If the prosecutor demands that the defendant be declared guilty, and the defender insists that the case be dropped: $b = \{a\}$, $d = 0$, the equation for the judge has no solution, i.e., the judge in a state of frustration. Thus, there is a case in which the judge does not choose alternative 1. Therefore, the trial is not perfect.

Appendix

Conclusion

1. The set of my alternatives is {1, 0}; 1 - to go, 0 - to stay home. The graph of relations, where a and b are my friends, is

Me

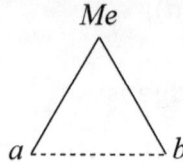

Fig. A2-26. Relations graph for problem 1

the diagonal form

$$[a] + [b]$$
$$[Me]\,[a + b]$$
$$[Me(a + b)] \quad ,$$

and the equation for me

$$Me = (a + b)Me + \overline{Me}.$$

My friends do not invite me: $a = 0$, $b = 0$, thus,

$$Me = \overline{Me}.$$

I am in frustration. I do not have a choice that would follow the anti-selfishness principle. This may serve as a stimulus to reconsider the situation. For example, perhaps I can invite my friends to visit me. In other words, I create a new pair of alternatives: invite friends for visit (1) – do not invite (0). The same equation corresponds to *Me*, but the values of a and b are different. I know that my friends want to see me, i.e., $a = 1$, $b = 1$. After substituting these values into equation, I find $Me = 1$. The model tells me that by choosing the alternative "to invite" I will act in accord with the anti-selfishness principle.

Bibliography

Batchelder, W. H. & Lefebvre, V. A. (1982).
A Mathematical Analysis of a Natural Class of Partitions of a Graph. *Journal of Mathematical Psychology*, No.4.

Campbell, D. T. (1997)
From evolutionary epistemology via selection theory to a sociology of scientific validity. *Evolution and Cognition*, No.3.

Kaiser, T. B., Schmidt, S. E. (2008).
Generalized Semantics for Reflexive Analysis of Groups. *Reflexive Processes and Control*, No.1.

Krylov, V. Yu. (2000).
Mathematicheskie i teoreticheskie problemy matematicheskoy psichologii (Mathematical and theoretical problems in mathematical psychology). Moscow: Yanus K.

Lefebvre, V. A. (1965).
Basic ideas of reflexive games logic, *Problemy Issledovania Sistem i Structur*, Moscow: AN USSR Press.

Lefebvre, V. A. (1966).
Elements of Logic of Reflexive Games, *Problemy Ingenernoy Psichologii*, No. 4.

Lefebvre, V. A. (1967).
Conflicting Structures, Higher Education Press, Moscow.

Lefebvre, V. A., Baranov, P. V., Lepsky, V. E. (1969).
Internal Currency in Reflexive Games, *Izvestia AN USSR, Tekhnicheskaya Kibernetika*, No. 4.

Lefebvre, V. A. (1982).
Algebra of Conscience. Holland: D. Reidel.

Lefebvre, V. A. (2001).
Algebra of Conscience (second enlarged edition). Holland: Kluwer.

Lefebvre, V. A. (2007).
A Reflexive Agent in a Group, *Reflexive Processes and Control*, No.1

Novikov, D. A., Chkartishvili, A. G. (2003).
Refleksivnye igry (Reflexive games). Moscow: SINTEG.

Schreider, Yu. A. (1999).
Continuously-valued logics Lef_m as languages of reflexion. *Nauchno-tekhnicheskaya informatsia*, No. 1-2.

Taran, T. A. (1998).
Model of Reflexive Behavior in a Conflict Situation. *Journal of Computer and Systems Sciences Int.*, No.1.

Taran, T. A. (2001).
   Many-valued Boolean Model of the Reflexive Agent. *Multi-Valued Logic*, No.7.
Taran, T. A., Shemaev, V. N. (2004).
   Boolean models of reflexive control and their application for describing informational warfare in social-economical systems. *Avtomatika i telemekhanika*, No.11.
Trudolubov, A. F. (1972).
   Decisions on dependency nets and reflexive polynomials. *VI Symposium po Kibernetike*, part III. Tbilisi.

www.ingramcontent.com/pod-product-compliance
Lightning Source LLC
Chambersburg PA
CBHW031809190326
41518CB00006B/262